Henrike Stephani

Automatic Segmentation and Clustering of Spectral Terahertz Data

Henrike Stephani

Automatic Segmentation and Clustering of Spectral Terahertz Data

Hyperspectral image analysis in the Terahertz domain

Südwestdeutscher Verlag für Hochschulschriften

Impressum / Imprint
Bibliografische Information der Deutschen Nationalbibliothek: Die Deutsche Nationalbibliothek verzeichnet diese Publikation in der Deutschen Nationalbibliografie; detaillierte bibliografische Daten sind im Internet über http://dnb.d-nb.de abrufbar.
Alle in diesem Buch genannten Marken und Produktnamen unterliegen warenzeichen-, marken- oder patentrechtlichem Schutz bzw. sind Warenzeichen oder eingetragene Warenzeichen der jeweiligen Inhaber. Die Wiedergabe von Marken, Produktnamen, Gebrauchsnamen, Handelsnamen, Warenbezeichnungen u.s.w. in diesem Werk berechtigt auch ohne besondere Kennzeichnung nicht zu der Annahme, dass solche Namen im Sinne der Warenzeichen- und Markenschutzgesetzgebung als frei zu betrachten wären und daher von jedermann benutzt werden dürften.

Bibliographic information published by the Deutsche Nationalbibliothek: The Deutsche Nationalbibliothek lists this publication in the Deutsche Nationalbibliografie; detailed bibliographic data are available in the Internet at http://dnb.d-nb.de.
Any brand names and product names mentioned in this book are subject to trademark, brand or patent protection and are trademarks or registered trademarks of their respective holders. The use of brand names, product names, common names, trade names, product descriptions etc. even without a particular marking in this works is in no way to be construed to mean that such names may be regarded as unrestricted in respect of trademark and brand protection legislation and could thus be used by anyone.

Coverbild / Cover image: www.ingimage.com

Verlag / Publisher:
Südwestdeutscher Verlag für Hochschulschriften
ist ein Imprint der / is a trademark of
AV Akademikerverlag GmbH & Co. KG
Heinrich-Böcking-Str. 6-8, 66121 Saarbrücken, Deutschland / Germany
Email: info@svh-verlag.de

Herstellung: siehe letzte Seite /
Printed at: see last page
ISBN: 978-3-8381-3490-1

Zugl. / Approved by: Linz, JKU, Diss., 2012; Kaiserslautern, TU, Diss., 2012

Copyright © 2012 AV Akademikerverlag GmbH & Co. KG
Alle Rechte vorbehalten. / All rights reserved. Saarbrücken 2012

Abstract

The goal of this thesis is to find ways to improve the analysis of hyperspectral Terahertz images. Although it would be desirable to have methods that can be applied on all spectral areas, this is impossible. Depending on the spectroscopic technique, the way the data is acquired differs as well as the characteristics that are to be detected. For these reasons, methods have to be developed or adapted to be especially suitable for the THz range and its applications. Among those are particularly the security sector and the pharmaceutical industry.

Due to the fact that in many applications the volume of spectra to be organized is high, manual data processing is difficult. Especially in hyperspectral imaging, the literature is concerned with various forms of data organization such as feature reduction and classification. In all these methods, the amount of necessary influence of the user should be minimized on the one hand and on the other hand the adaption to the specific application should be maximized.

Therefore, this work aims at automatically segmenting or clustering THz-TDS data. To achieve this, we propose a course of action that makes the methods adaptable to different kinds of measurements and applications. State of the art methods will be analyzed and supplemented where necessary, improvements and new methods will be proposed. This course of action includes preprocessing methods to make the data comparable. Furthermore, feature reduction that represents chemical content in about 20 channels instead of the initial hundreds will be presented. Finally the data will be segmented by efficient hierarchical clustering schemes. Various application examples will be shown.

Further work should include a final classification of the detected segments. It is not discussed here as it strongly depends on specific applications.

Contents

1 Introduction   **1**
   1.1 Spectroscopic Techniques for the Detection of Chemicals   3
   1.2 Contribution   6

2 Preprocessing   **11**
   2.1 Standard Methods   12
   2.2 Time-Domain   15
   2.3 Frequency-Domain   25
   2.4 Conclusion   36

3 Simulation of THz-TDS Spectra   **37**
   3.1 Basic Shape   38
   3.2 Noise   43
   3.3 Peaks   46
   3.4 Conclusion   49

4 Feature Reduction   **51**
   4.1 State of The Art   52
   4.2 Representing the Magnitude Spectrum   56
   4.3 Correlation Based Evaluation   67
   4.4 Additional Signal-Features   74
   4.5 Conclusion   77

5 Automatic Segmentation of Spectral Data   **79**
   5.1 State of the Art   80
   5.2 Clustering Model   81
   5.3 Distances in Classical Hierarchical Clustering   82
   5.4 Discussion of Karypis Chameleon Algorithm   90
   5.5 Simplified Chameleon   96
   5.6 Spatial Domain   98

Contents

| | 5.7 | Evaluation of Hierarchical Clustering | 103 |
| | 5.8 | Conclusion | 109 |

6 Application Examples   **113**
    6.1 Eight Chemical Compounds 115
    6.2 Lactose Pellet 127
    6.3 Mockup Mail Bomb 131
    6.4 Application to Raman Spectra 143
    6.5 Conclusion 146

7 Conclusion   **149**
    7.1 Future Research 151

Implementations   **153**

Parts of this Work Previously Published In   **157**

Bibliography   **159**

1 Introduction

In classical image processing, objects are analyzed by their representation in the visible domain, i.e. by the way they appear to a human eye. This is done either by one gray value or three color values. However, there are spectral ranges other than the visible ones that can be used for object identification. In the area of spectroscopy the interaction between an object and an emitted radiation of specific wavelengths is analyzed. This interaction can then be used to unravel characteristics of the measured material that cannot be detected in other ways.

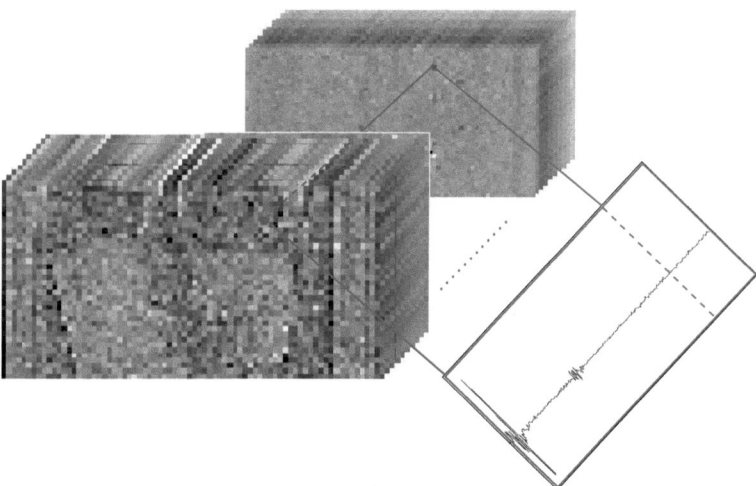

Figure 1.1: Schematic view of a hyperspectral image.

1 Introduction

Traditionally spectroscopy is used to acquire single spectra. However, over the last years, measuring spectra that represent larger spatial areas has become more and more popular. In this field, images are measured that contain hundreds of values in each pixel representing their spectral characteristics. They are therefore called hyperspectral images. The problem that arises is how to represent this type of data. In Figure 1.1 a hyperspectral image is illustrated. One can see that, although it can be provided in a three-dimensional array, it should not be mistaken for a volume image. Usually, hyperspectral images do not contain any information about a third spatial dimension. A volume rendering therefore, does not help in the analysis.

To illustrate its content, we depict one example pixel in red in Figure 1.1 together with its spectrum. Next to regarding the spectra separately like this, another possibility is to show a stack of two-dimensional images, each one standing for one measured channel (i.e. sampled time or frequency step). In the figure, the first and the last 10 channels are shown. However, the complete stack in Figure 1.1 consists of 3200 images, not just 20. Analyzing them separately is hardly feasible.

Image 1.1 is acquired by Terahertz time-domain spectroscopy (THz-TDS), which is a relatively new technique and not well researched. Of course, hyperspectral imaging in general is not limited to the THz domain. Many spectroscopic techniques can be used. And there already is a vast amount or literature on the subject. However, much of this literature focuses on near-infrared applications, as it is particularly established in real-world scenarios (esp. remote sensing)[6, 7], some focus on the medical sector with magnet resonance tomography, and mass spectroscopy [8, 9]. Only little is concerned with hyperspectral THz-TDS imaging.

Therefore, the goal of this thesis is to find ways to improve the analysis of hyperspectral Terahertz images. Although it would be desirable to have a solution that works independently from the special spectroscopic technique, this is impossible. The way the data is acquired differs as well as the characteristics that are to be detected. For these reasons methods have to be developed or adapted which are especially suitable for the THz range and its applications. Among those are particularly the security sector and the pharmaceutical industry. We aim at developing methods that preserve and improve as much of the spectral content as possible on the one hand. And on the other hand we want to give a representation that can easily be used to detect to content of interest with as little manual interference as possible.

1.1 Spectroscopic Techniques for the Detection of Chemicals

Before we go into a more detailed description of the contributions of this thesis, we will give the reader a short overview of traditional spectroscopic techniques and their advantages and drawbacks. We will then explain the technique of THz time-domain spectroscopy which will be used throughout this thesis.

1.1 Spectroscopic Techniques for the Detection of Chemicals

The electromagnetic spectrum with its different frequency regions is shown schematically in Figure 1.2. There is a wide range of spectroscopic techniques that are used to cover these regions. Apart from visible range photography, among the well known ones are near-infrared and mid-infrared spectroscopy, spectra measured with the photo-spectrometer, nuclear magnetic resonance images, Raman spectroscopy and for about 20 years now Terahertz spectroscopy. Applications range from astronomy and the inspection of paintings to remote sensing of landscape, they include medical analysis of body tissue and pharmaceutical analysis, i.e. biochemical applications [8].

We will focus on applications that involve the detection of chemicals. Therefore we will introduce the reader to some of the other wavelength that are used in this area. Subsequently we give a more detailed introduction into applications and technologies that focus on the Terahertz range.

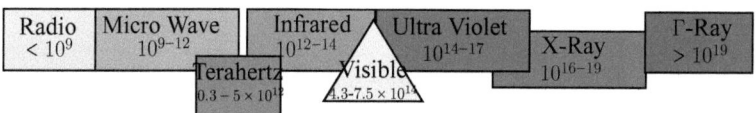

Figure 1.2: Spectral regions of electro magnetic spectrum with respective frequencies [10].

1 Introduction

1.1.1 Raman, Near-Infrared, and Mid-Infrared

Electro magnetic waves of each wavelength range unveil specific characteristics that can be used to identify materials. If the aim is to identify chemical compounds, the techniques that are mostly used are Raman, near-infrared, and mid-infrared spectroscopy. To embed the THz range into these state of the art methods, we will give a short overview. It will consist of an outline of the properties, advantages and disadvantages.

Raman spectroscopy: This technique is based on the so-called Raman effect. It is caused by inelastic scattering of monochromatic incident light, i.e. the light of one frequency is scattered and produces energy at a slightly shifted frequency. The incident light usually comes from the near-infrared, visible or ultraviolet range. The scattering is caused by molecular bond vibrations, thereby the respective bonds can be identified. Raman spectroscopy can thus be used to identify chemicals. The advantages of Raman spectroscopy are that little sample preparation is needed and water marks no problem. Glass and polymers usually do not hinder the measurement. The downside is that to obtain a Raman spectrum only the inelastic scattering is accounted for while the main radiation that passes through the sample is suppressed. Hence it has a nature of being a "secondary radiation", it is weak and the apparatus needs to be optimized. Furthermore, other effects such as autofluorescence can hide the radiation altogether. Additionally Raman spectroscopy can be destructive due to a high laser intensity [11, 12].

Near-Infrared (NIR) spectroscopy: The phenomenon that is observed here, is the absorption of energy at specific frequency bands of incident light. The absorption is caused by molecular overtone and combination vibrations. The spectra are complex because of their relatively low intensity and the broad overlapping bands. NIR is not sensitive to water which is one of the reasons that NIR reflectance spectroscopy is mainly used in the food and agricultural industry [13, 14]. Especially in remote sensing of soils NIR is applied. The applications in the pharmaceutical industry are more recent because of the complexity. Generally, the advantages of NIR are that little sample preparation is required, it is very quick, glass is non-absorbing and physical as well as chemical parameters can be extracted. Disadvantages are that sophisticated calibration has to be performed to extract them and the absorption is very broad and overlapping [15, 16].

1.1 Spectroscopic Techniques for the Detection of Chemicals

Mid-Infrared (MIR) spectroscopy: One can say that both near- and mid-infrared are complementary to Raman spectroscopy. Like near-infrared spectroscopy MIR observes the absorption at different frequency bands. Vibrational modes as well as some rotational ones can be detected. Advantages are that the absorption lines are usually very fine and single compounds can be identified by analyzing the acquired spectra. It is the method that is most popular in analyzing chemical compounds. An extensive library containing the respective absorption bands is publicly usable [17]. Quantitative as well as qualitative analysis of the spectra is performed. Disadvantages are that sample preparation is need, the method can be destructive and glass is absorbing, hence, no glass containers can be used [18]. Furthermore, this region is sensitive to water.

In addition to the advantages and disadvantages of the addressed methods it has to be said, that the question of "which method to use" highly depends on the material that is to be analyzed and the environmental conditions during the analysis. Remote sensing of soils for example is not possible when the radiation is as weak as the Raman effect or the method is as sensitive to water as MIR. On the other hand the disadvantages of MIR, such as sample preparation, higher costs, and longer acquisition duration are minor in many highly controlled and expensive settings in the pharmaceutical industry, whereas the reliability of the results, i.e. the correct classification is essential.

Therefore, these methods often do not directly compete against each other but rather complement each other. In this context we will now introduce the radiation that builds the focus of this work, the Terahertz radiation. In many aspects it is complementary to the other techniques mentioned above and its novelty is the main reason for its rarity in industrial applications.

1.1.2 Terahertz Time-Domain Spectroscopy (THz-TDS)

In Terahertz time-domain spectroscopy (THz-TDS), similar to MIR and NIR the absorption of chemicals is analyzed. Absorption lines are generated by inter- and intra-molecular rotational modes within the substrate to be measured. In contrast to infrared spectroscopy one can differentiate between polymorphous shapes of a compound. For this reason different forms of Aspirin for instance can be distinguished in the THz domain [19]. When measuring solids in the THz wavelength, the phonon, i.e. collective excitations are dominant. Especially crystalline structures can be identified.

One advantage of THz-TDS for pharmaceutical quality control is that amor-

1 Introduction

phous material — which for example coating of drugs often consists of — is non-absorbing in all the three frequency ranges of NIR, MIR, and THz [20]. One of the biggest advantages of THz-TDS is, however, that additionally most packaging materials are non-absorbing. This includes carton, plastics, paper, ceramics, but also clothes. Therefore, this range can be used for many purposes. Among those are biological, medical or pharmaceutical applications, applications within the security sector and non-destructive testing [21]. The reason why the THz bandwidth has not been exploited yet, is that for a long time there has been a gap between the radio and the optical frequencies. The transistor technologies could cover frequencies up to 50 GHz while the laser technologies could be extended down to 30 THz [22]. With the development of ultra-short laser pulse emitters and detectors in the early 90s and the further technological development in the last two decades this gap closes more and more [23].

There are different ways how to acquire spectra in the THz range. We will focus here on Terahertz Time-Domain spectroscopy (THz-TDS) also called Terahertz pulsed spectroscopy. An ultra-short femtosecond laser pulse is generated and sent through an object. The detector then measures the reflection by or the transmission through the object. This technique has a very broad bandwidth, often covering several THz, and can therefore be used to detect highly different features of unknown substrates without knowing their exact position beforehand — in contrast to frequency resolved techniques where usually only some frequency bands are analyzed. One further advantage is that it is insensitive to thermal background noise [24] because it is time resolved. Hence, measurements can be taken at room temperature. This and the fact that packaging materials are non-absorbing make it applicable in areas where infrared spectroscopy for example can not be used.

At the beginning of every chapter in this thesis we will outline the content of the respective chapter. Within this outline the contributions will be written in italic fonts.

1.2 Contribution

Due to the fact that in many applications the volume of spectra to be organized is high, manual data processing is difficult. Especially in hyperspectral imaging, the literature is concerned with various forms of data organization such as feature reduction and classification [8, 9]. In all these methods, the amount of necessary

1.2 Contribution

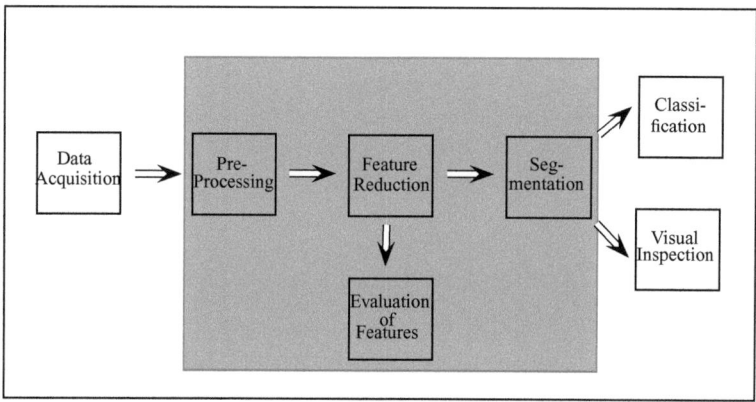

Figure 1.3: Different steps of spectral data processing. Green region includes the different topics of this thesis.

influence of the user should be minimized on the one hand and on the other hand the adaption to the specific application should be maximized.

The goal of this thesis is to automatically segment or cluster THz-TDS data. To achieve this goal, we propose a course of action depicted in Figure 1.3. By considering these steps one can design methods adaptable to different kinds of measurements and applications. We will focus on the green steps. That means we assume to have no influence on the data acquisition and we will not discuss a final classification, as the latter strongly depends on specific applications. The green boxed course of action also builds the red line of the thesis. In the following chapters we will go into detail for all of the steps. state of the art methods will be analyzed and supplemented where necessary, improvements and new methods will be proposed. Some of the algorithms do not need to be applied on all measurements but only if a certain application or data acquisition quality makes them necessary.

Chapter 2: Different preprocessing step are discussed. Because the THz-TDS technology is rather new, the standardization is not high and often only very little preprocessing is applied at all. We therefore propose to add some algorithms and procedures to improve the data quality. These additions result in various

1 Introduction

techniques. In Section 2.2 two methods to reduce the noise in the time-domain are explained. Preprocessing in the frequency-domain is discussed in Section 2.3. Here, we first discuss existing approaches that can be used for peak preserving smoothing and then introduce methods to determine the dynamic range of a THz-TDS measurements and to estimate a correct baseline.

Chapter 3: We propose a procedure to simulate spectra that have the same basic shape, noise, and peak characteristic as THz-TDS absorption spectra. Two different kinds of approaches on basic shape simulation from the literature are compared with real-world measurements in Section 3.1. Noise characteristics are analyzed and included in Section 3.2. Finally the typical peak shape and possible positions that can be found are inserted by a spline based method in Section 3.3.

Chapter 4: Possibilities for feature reduction of THz-TDS spectra are analyzed. The main features lie in the magnitude spectrum, therefore the focus is on them. In Section 4.2 we start by discussing one of the state of the art methods from that area, namely PCA. We then propose two other feature sets. One is based on real wavelets one on complex wavelets. In both cases the dimensionality of the spectra is reduced while the representation remains spectrally comprehensive and interpretable.

Both methods are evaluated with the simulation scheme from Chapter 3. The evaluation in Section 4.3 is applied to compare different sets of basis functions and support sizes for real wavelets, i.e. finding the best parameters. Furthermore, it is used to compare the proposed feature sets with each other as well as the given full spectrum. In the end of the chapter standard features as well as a phase representation are discussed and finally all features are combined in Excursus 2 into a fixed set of features that can be used in various applications.

Chapter 5: Initially, clustering methods in general are discussed to then focus on hierarchical clustering. In Section 5.3, after explaining the classical algorithm, we discuss different parameters that need to be set and propose to use an atypical distance measure to improve the results. For high data volume the classical hierarchical clustering easily fails, therefore the Chameleon algorithm by Karypis is used. We discuss its functionality in Section 5.4 and discuss problems with the initially incorporated similarities of the algorithm and propose an alteration. In Section 5.5 we propose an alternative but similar approach to high volume hierarchical clustering. It is based on the Chameleon as well as the classical hierarchical clustering.

The data organization up to this point is done wholly based on the information of the single spectra and can be applied to independent measurements as well as images. In the next Section 5.6 we therefore investigate two approaches on including spatial information into the segmentation of hyperspectral images. The first one consists of applying classical image processing on the feature set proposed in the previous chapter and apply the segmentation on the pre-smoothed data. The other one is an approach that includes a spatial similarity directly into the clustering by combining a spectral with a spatial distance measure. Section 5.7 finally discusses the cluster evaluation. Problems with the state of the art method in automatic approaches are explained. We propose an alternative approach and particularly an algorithm for its implementation.

Chapter 6: We present various applications of the methods and algorithms proposed in this work and thus illustrate their usefulness. We use high dynamic range measurements of chemicals to illustrate the spectral preprocessing as well as to practically evaluate the parameter choice in hierarchical clustering for these kinds of data in Section 6.1. A hyperspectral image of a Lactose pellet shows the usability of the time-domain preprocessing in Section 6.2. A relatively big hyperspectral measurement of an envelope with different materials inside serves to illustrate and compare the feature sets as well as the high volume clustering algorithms in Section 6.3.

Finally we also apply the feature reduction and the data clustering on different kinds of spectra measured by Raman spectroscopy to illustrate the potential for other areas in Section 6.4.

i

1 Introduction

2 Preprocessing

As pointed out in the introduction, Terahertz time-domain spectroscopy (THz-TDS) is a time-resolved technique. That means an ultra-short laser pulse acquired in the time-domain leads to a relatively wide spectrum in the frequency-domain. Within this frequency-domain one can then potentially find characteristics, that are used to identify a material. However, before automatic classification or organization of this data can begin, often a number of preprocessing steps have to be applied to make the data comparable and to eliminate artifacts. Preprocessing of THz-TDS data will hence be the topic of this chapter.

Contribution *Initially the standard procedure that is used to get from the time- to the frequency-domain will be explained. Beforehand, one needs to consider the appearance of echo pulses in the time-domain. To eliminate these echo pulses we apply two alternative methods, based on windowing or wavelet shrinkage. They are explained in Sections 2.2.1 and 2.2.2. Once transformed into the frequency-domain usually the following issues have to be considered. The smoothness, the dynamic range and the non-constant baseline. In more detail, the spectra might not be smooth enough, i.e. oscillations that signify no characteristic absorption might be present, for this purpose we compare two different peak preserving smoothing filters in Section 2.3.1. To determine where the noise begins in a spectrum, i.e. the dynamic range ends, we propose an algorithm in Section 2.3.2. Furthermore, to estimate the baseline of a measurement we propose to simulate them based on their physical origin in Section 2.3.3.*

Before explaining the details of the proposed preprocessing scheme, two remarks have to be made as to the specific form of the measured compounds:

Remark 2.1 We will focus on transmittance spectra where a sample signal and a reference signal are acquired. Most preprocessing steps can be used for reflection spectra as well only without the existence of a reference measurement.

Remark 2.2 All measurements considered in this thesis are measured in dry air. Terahertz radiation is very sensitive to high humidity in the air. Figure

2 Preprocessing

2.1 shows the difference between a reference with and without water absorption lines. The reference is a measurement taken without an object between emitter and detector.

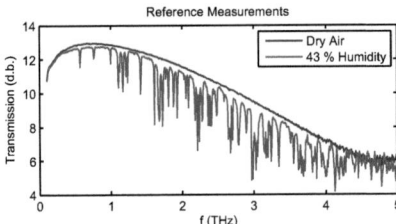

Figure 2.1: Reference spectrum with and without water absorption lines.

Although the exact positions of the waterlines are known, they are difficult to eliminate. Especially in higher frequent regions (above 1.5 THz) their density is high, this can make the detection of real peaks in a sample spectrum taken in humid surroundings difficult. Eliminating these water lines from the spectrum is a separate research topic and is not within the scope of this thesis. However, the known positions of these waterlines (e.g. in the HITRAN database [25]) can be used for their elimination as done by [26] for example.

2.1 Standard Methods

To generate THz spectra with a broad bandwidth, an ultra-short laser pulse is emitted and its transmittance through an object or reflection by an object is measured in the time-domain signal. Information derived from the frequency-domain, namely the absorption coefficient and the refractive index, is then used to determine the characteristics of the objects. We will now explain the classical procedure how to calculate these indicators from a transmission (2.1) measurement.

(1) The time-domain signal of the sample and the reference is acquired. In Figure 2.2 two typical THz-TDS signals are displayed. The blue signal is the reference pulse and the green one the sample. The sample pulse is time-delayed, has a reduced main amplitude, and shows deformations after the main amplitude.

2.1 Standard Methods

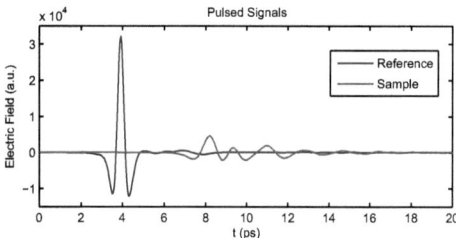

Figure 2.2: Time-domain signal of sample and reference.

(2) For both signals the Fourier transform is calculated. The Fourier transform $\hat{S}(\nu)$ of a signal $s(j)$ with $j \in \{0, ..., n-1\}$ is defined by:

$$\hat{S}(\nu) = \sum_{j=0}^{n-1} s(j) e^{-2\pi i \frac{\nu j}{n}}.$$

The left hand side of Figure 2.3 shows the logarithmic magnitude spectra of both measurements while the right hand side shows the unwrapped phase spectra.

(3) Out of the two complex Fourier transforms the following coefficients are calculated:

(a) The frequency dependent index of refraction:

$$n(\nu) := 1 + \frac{c}{2\pi \nu d} \Phi(\nu)$$

where d is the thickness of the sample and $\Phi(\nu)$ its phase spectrum.

(b) The absorption coefficient:

$$\alpha(\nu) := -\frac{2}{d} \ln \left\{ A(\nu) \frac{[n(\nu)+1]^2}{4n(\nu)} \right\} \quad (2.1)$$

where d is the thickness of the sample, $n(\nu)$ its refractive index, and $A(\nu)$ the magnitude spectrum.

13

2 Preprocessing

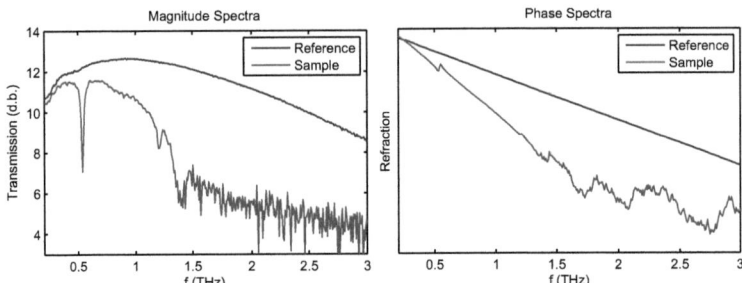

Figure 2.3: Left: Logarithmic magnitude spectra of sample and reference; Right: Phase spectra.

(4) A type of normalization of the spectrum is achieved by deconvolving the sample signal with the reference. This deconvolution in the time-domain can be executed as a division in the Fourier domain. Therefore, $A(\nu)$ and $\Phi(\nu)$ are not the magnitude and phase, respectively, of the sample alone but rather of the ratio of the samples and reference spectra [27]. The phase of a THz-TDS measurement is almost linear, due to the single cycle pulse measured [28]. Nevertheless, the index of refraction is often even assumed to be locally constant [27] with a constant sample thickness. If this is the case the absorption coefficient does only depend on the ratio of the two magnitude spectra and the other parameters only have a contribution of a constant factor.

We are mainly interested in the overall absorption as well as the peak positions of the spectra. In Figure 2.3 the linear shape of the phase is visible. The characteristic peaks are not very dominant while in the magnitude they are very pronounced. Therefore, we focus on the magnitude spectra of the ratio between sample and reference as a representation of the absorption coefficients. Figure 2.4 illustrates the transmittance resulting from these calculations.

These are the standard steps that are carried out to get an interpretable spectrum. As the measurements differ due to the system parameters, environmental conditions, and the measuring person, further preprocessing is necessary. In the following sections we will talk about the means we propose to enhance the qual-

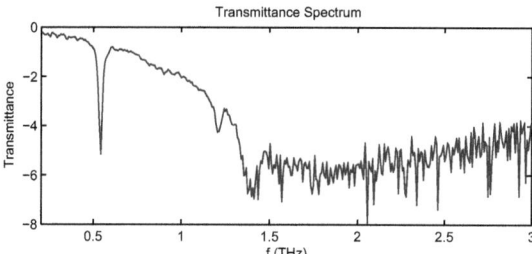

Figure 2.4: Logarithmic sample magnitude divided by reference.

ity of the spectra in general and to deal with specific effects. The overall goal of this preprocessing is to gain spectra that are adjusted for further automatic interpretation and classification.

2.2 Time-Domain

To avoid carrying artifacts through the calculation process, they have to be eliminated as soon as possible. Therefore, the first issue in preprocessing THz-TDS spectra is to remove effects that can already be detected in the time-domain. The most important of these effects is the appearance of echo pulses.

We assume that every measurement contains at least one pulse, the maximum. With the exception of information in direct vicinity of the main pulse — i.e. 2 picoseconds before and 10 picoseconds after it — we consider every maximum that exceeds the standard deviation of the remaining signal to be an echo pulse. We call all pulses except the main pulse echo pulses — even if they appear before the main pulse and are not strictly speaking echoes. Additional pulses appear for mainly two reasons. Either the broad focus of the laser beam or reflections within the sample. Both will be explained in more detail later on.

Although some of these pulses can be used for further analysis they still have to be detected and separated from each other. Otherwise they lead to strong interferences within the spectral domain. We will first present our approach on filtering out pulses that occur before the main pulse and that are relatively easy to filter out. Then we will talk about echos behind the main pulse.

2 Preprocessing

2.2.1 Echo Removal with Window Functions

In THz-TDS the lateral resolution of imaging systems is low. Therefore, the diameter of the focus beam can be bigger than the object itself. This may produce two pulses reaching the detector; one of them being the usual sample pulse that went through the material. The other one, however, is a remainder of the reference, i.e. the original pulse [29]. As this reference remainder did not have to pass through the sample, the time delay is almost zero in contrast to the sample signal's time delay. Hence, it appears before the main sample's pulse. Generally one can assume everything appearing before that pulse to be noise and filter it completely away, for that purpose we apply apodization functions.

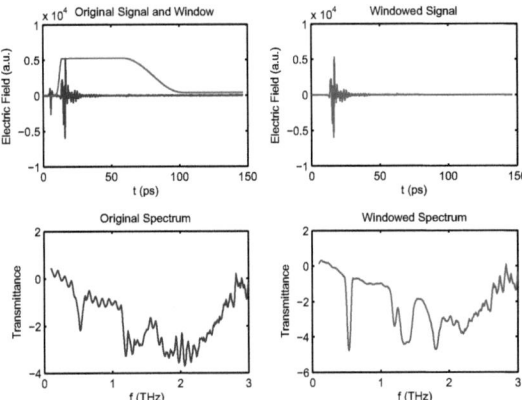

Figure 2.5: Top Left: Signal with two pulses; Top Right: Signal filtered, such that only the main pulse remains; Bottom Left: Additional echo pulse produces interferences in spectrum; Bottom Right: Suppression of additional pulse in time-domain pulse, suppresses interferences in the spectrum.

2.2 Time-Domain

Apodization or window functions are very often used prior to the Fourier transform to eliminate undesirable aliasing effects [30]. Aliasing appears because the Fourier transform acts under the assumption of infinite periodic signals while experimentally given data usually is finite. In Figure 2.5 the signal to noise ratio (SNR) is so high, that this effect of so-called spectral leakage should not be dominant in the Fourier spectrum. However, not all measurements have such a good SNR. Therefore, we apply apodization for classical reasons — i.e. behind the main pulse — as well as to remove remainders of the reference pulse before the main pulse.

We apply two kinds of windows to deal with the echoes before the main pulse on the one hand and the spectral leakage on the other hand. Information before the main pulse is suppressed generously with a Nuttall window [31] while the one coming after the pulse is hardly altered. Only to eliminate the effects of spectral leakage a Hamming window is used [32]. Figure 2.5 shows the importance of the elimination of the reference pulse. The displayed spectrum has its main peaks at 0.5 THz and between 1.3 and 1.6 THz. Within the blue spectrum on the left hand side the interferences originating from the extra pulse can be seen. In the red spectrum the peaks are well preserved while the interferences are filtered away.

There are echo pulses that are more difficult to filter. For this purpose we will apply a form of wavelet shrinkage in the next section. The wavelet transform and its useful characteristics will be applied at various points of this thesis. Therefore, we will now give a short Excursus about the most important ideas of this well known transform.

1 Excursus (The Discrete Wavelet Transform)
The wavelet transform has the advantage of extracting information of specific frequency characteristics and in addition preserving the lateral position of these characteristics. Furthermore, it is especially appropriate when the signals are only finitely oscillating, i.e. substantially differ from sinusoidals. To facilitate understanding wavelet based methods that are applied in this thesis, we now explain the concept of the discrete wavelet transform. However, as this is a well-known topic and not the focus of the work we refer the reader to the literature as for example [33, 34].

The basis of the discrete wavelet transform is a multiscale analysis. One can roughly say that a multiscale analysis is a sequence of subsets $(V_j)_{j \in \mathbb{Z}}$ of $L^2(\mathbb{R})$ (satisfying a number of conditions, see for example [35]) that can be used to

2 Preprocessing

represent characteristics of functions $u \in L^2(\mathbb{R})$ up to a certain coarseness level j. For this purpose the projections of the functions to the subsets V_j are used, i.e. $P_{V_j}u$. Additionally one can define for each V_j the orthogonal complement W_j. A function can then be represented up to coarseness j by the projection $P_{V_j}u$ and the left over information represented by using the W_i. Following from this, one can represent u by using W_i:

$$u = \sum_{i \in \mathbb{Z}} P_{W_i} u = P_{V_j} u + \sum_{i \geq j} P_{W_i} u \qquad (2.2)$$

Such a multiscale analysis is generated by a function $\Phi \in V_0$. This function is defined as $\Phi_{j,k}(x) := 2^{-j/2}\Phi(2^{-j}x - k)$ with $k \in \mathbb{Z}$ are orthonormal basis of V_j. The function Φ is called the generator or scaling function of the sequence. On the basis of such a scaling function one can also define a basis $\{\psi_{j,k} | k \in \mathbb{Z}\}$ for the W_j in the following way:

$$\psi_{j,k}(x) := 2^{-j/2}\psi(2^{-j}x - k) \qquad (2.3)$$

with

$$\psi(x) := \sqrt{2}\sum_{k \in \mathbb{Z}}(-1)^k h_{1-k}\Phi(2x - k), \text{ where } h_k := (\Phi, \Phi_{-1,k}).$$

Following from Equation 2.2 we can use the basis for the W_j to find a basis for $L^2(\mathbb{R})$, namely $\{\psi_{j,k} | j, k \in \mathbb{Z}\}$. Now, using Equation 2.3 in Equation 2.2, one can represent $u \in L^2(\mathbb{R})$ by $u = \sum_{j,k \in \mathbb{Z}} (u, \psi_{j,k}) \psi_{j,k}$. The functions ψ are called wavelets.

Although it is not trivial to find a generator function for a multiscale analysis, a multitude of wavelets and respective generator functions have been developed in the last decades.

One of the computationally most attractive characteristics of the wavelet coefficients is that they can be recursively calculated by filtering the signal at each level with a low-pass filter $h_\Phi(l) := h_l$ and high-pass filter $h_\psi(l) := (-1)^l h_{1-l}$. For the calculation of the wavelet coefficients $\gamma_\psi(j,k) := (u, \psi_{j,k})$ and the scaling function coefficients $\gamma_\Phi(j,k) := (u, \Phi_{j,k})$ one can prove that the following two equations hold:

$$\gamma_\psi(j,k) = \sum_{l \in \mathbb{Z}} h_\psi(l - 2k)\gamma_\Phi(j-1, l) \qquad (2.4)$$

and
$$\gamma_\Phi(j,k) = \sum_{l\in\mathbb{Z}} h_\Phi(l-2k)\gamma_\Phi(j-1,l). \qquad (2.5)$$

That means at a fixed coarseness level j the coefficients of both scaling and wavelet function can be calculated on the basis of the scaling function coefficients of the previous level $j-1$.

Explained in a more intuitive manner, one can imagine the signal to be split up at each level into a low-pass and high-pass component, i.e. into coarse and fine information of a certain level. The low-pass component is then used for the calculations of the next fine level. This splitting also leads to a dyadic downscaling of the coefficients in each step. The reconstruction of a wavelet decomposed signal can be performed analogously by up-sampling and filtering with adjoined filter-operators.

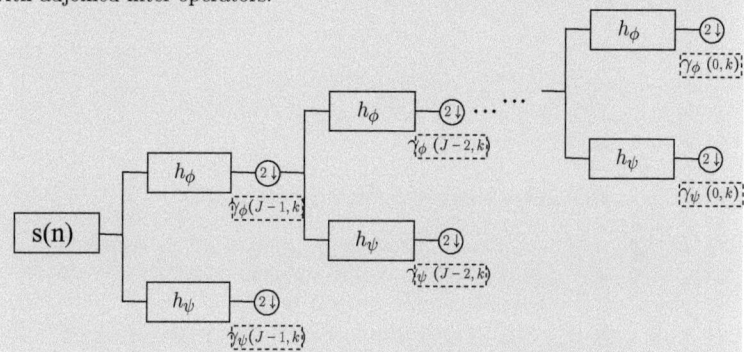

Figure 2.6: Cascade algorithm for calculation of discrete wavelet coefficients for signal $s(n)$ with $n = 0, ..., N-1$ and $N = 2^J$. Low-pass filter h_ϕ and high-pass filters h_ψ are used to recursively calculate each level js scaling function coefficients $\gamma_\phi(j,k)$ and wavelet coefficients $\gamma_\psi(j,k)$ on the basis of the previous scaling function coefficients $\gamma_\phi(j+1,k)$ in accordance with Equation 2.4 and 2.5.

Note that for finite signals $s(n)$ with $n = 1, ..., N$ (and for the dyadic downsampling $N = 2^J$) the sums in Equation 2.4 and 2.5 are finite as well. The downsampling and filtering scheme for a finite signal is illustrated in Figure 2.6. The algorithm is started by first applying the filters on the initial signal

2 Preprocessing

$s(n)$. Thereby the initial level of wavelet as well as scaling function coefficients is retrieved. With a sampling size of 2^{J-1} the initial level is called the $J-1$th level as it contains 2^{J-1} coefficients. In each recursion only the previous scaling function coefficients have to be known. They are used as the basis for the further calculation of coefficients as denoted in Equations 2.4 and 2.5.

In the finite case one has to consider that the filter has to evaluate undefined regions at the borders. In this work we apply the classical approach of assuming a periodically continued signal.

With this downsampling scheme and if the wavelet basis function has a compact support, the calculation of wavelet coefficients only has linear time complexity $O(N)$, with N being the number of samples. Furthermore, if only high frequent information is of interest, e.g. for denoising purposes, the downsampling scheme does not have to be carried out until the 0th scaling level but rather can terminate earlier.

For all calculations in this work we have used the Wavelab package by Donoho et al [36].

2.2.2 Echo Removal with Wavelet Shrinkage

The other source for additional pulses in the time-domain are internal reflections during the measurement. As mentioned before, these reflections are not wholly undesired. They are used to determine material parameters such as the thickness of different layers [37, 38]. The appearance of several pulses is, however, not useful when it comes to determining the spectral characteristics of the material, such as the absorption lines that originate from the polymorphous shape of the material. As already shown in Figure 2.5 these additional pulses also lead to interferences in the spectrum [19]. Hence, they have to be separated from the original signal.

Filtering away information before the main pulse is relatively easy as there should be no information of interest. Unfortunately, with real echos this is not the case. We therefore have to apply a method that detects only the pulse information and separates it from all surrounding information that can be useful for spectral characteristics.

In Figure 2.7 we show a signal that has its first echo pulse at about 32 picoseconds (ps) and at 95 ps another one. One can note that the two echo peaks

2.2 Time-Domain

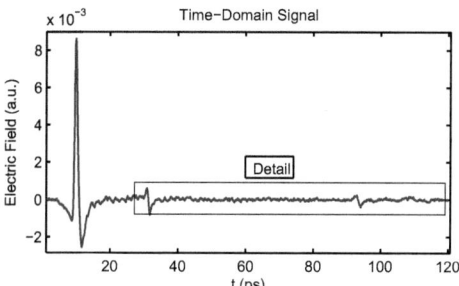

Figure 2.7: Signal with two echo pulses. Detail will be used to illustrate wavelet shrinkage echo removal in Fig. 2.8.

have a decisively different spectral characteristic than the surrounding information. This is even more obvious in the detailed view on the top left of Figure 2.8. The echos are broader and more dominant. Hence, to detect these peaks and thereby be able to separate them from the rest of the signal using a frequency based filtering method suggests itself. We have to use a method that provides time information as well, as the main pulse has the same spectral characteristics as the echo pulses but should not be filtered out. For these reasons we propose to use wavelet shrinkage because it is a technique that permits frequency as well as time dependent filtering.

By calculating wavelet coefficients as described in Excursus 1 we can extract a representation of the data in different frequency levels. This representation is then used to apply filtering on those scaling levels that contain information with the frequencies of interest. This level dependent coefficient filtering is called wavelet shrinkage. We follow the description of the different steps of this shrinkage procedure given by [39]:

Assuming a given dataset $X(t)$ consist of the signal of interest $R(t)$ and a noise component $N(t)$, i.e. $X(t) = R(t) + N(t)$, the shrinkage is usually applied as described in Algorithm 2.1. Our noise model is a little bit different, though. We consider the echos to be noise and want to keep the information that lies beneath the threshold, i.e. the noise $N(t)$, and filter out everything above it. Nevertheless, we will keep the introduced notation, where the noisy data will be the detail $X(t)$ of the original signal $S(t)$. The detail $X(t)$ contains the echos

2 Preprocessing

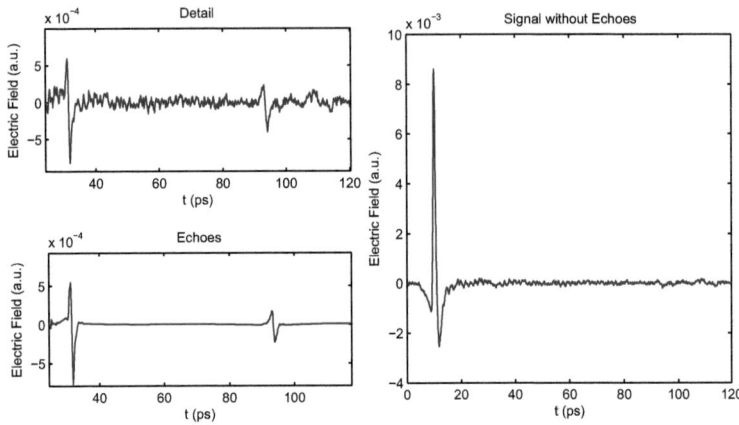

Figure 2.8: Top Left: Detail of Figure 2.7 with two echo pulses; Bottom Left: Result of the wavelet shrinkage with only echo pulses; Right: Signal without echo pulses.

$R(t)$ and additive "noise" $N(t)$. The result of the shrinkage will then be the approximation of the echoes $\hat{R}$ and what we will use in the end is not $\hat{R}$ (the pure echo pulses) but rather the "noise" given by $X(t)\hat{R}(t)$.

As already mentioned the denoising is not applied on the full signal, because we do not want to filter out the main pulse. We rather assume the first 10 picoseconds after and 5 picoseconds before the main pulse to contain information of interest that is allowed to have similar spectral characteristics as the main pulse itself. Everything beyond, having these kind of characteristics, is considered to be an echo. The signal to be denoised consists in the two parts before and after the main pulse. On both of them the shrinkage as described in Algorithm 2.1 is applied but naturally $N(t)$ is not white noise. Hence, applying a global threshold will not extract the echoes $\hat{R}$. We therefore apply a hard but level based threshold. In each coefficient level we calculated θ separately. For this, the variance in each level is used and everything lying beneath it is filtered out. Thereby only the echo pulses remain.

The whole procedure of this wavelet shrinkage echo removal is described in

2.2 Time-Domain

Algorithm 2.1: Classical Wavelet Shrinkage

Data: Noisy data $X(t)$
Result: Estimation of denoised data $\hat{R}(t)$
for $X(t)$ **do**

　Calculate wavelet transform $Y = W(X)$
　Calculate thresholds $\theta = d(Y)$, adequately chosen for the respective noise model for wavelet coefficients (Choice of d, see literature, e.g. [39]
　In each level remove information lying below the calculated thresholds — according to chosen thresholding method, such as hard, soft or fuzzy thresholding — and gain filtered transform: $Z = D(Y, \theta)$
　Recalculate signal with inverse wavelet transform of thresholded coefficients: $\hat{R} = \mathscr{W}^1(Z)$

Algorithm 2.2. Naturally $N(t)$ is not white noise, so applying a global threshold will not extract the echos $\hat{S}(t)$. On the bottom left of Figure 2.8 the filtered echo pulses can be seen. In the next step we subtract these echo pulses from the original signal and thereby get the denoised signal that is shown on the right hand side of this figure.

The positive effect of this echo removal on the spectral range is shown in Figure 2.9. In Figure 2.8 is illustrated that the echo pulses seem to be well extracted from the rest of the signal in the time-domain. In the frequency-domain the interferences that appear on the left hand side of the figure are mostly suppressed on the right hand side. At the same time sharp peaks or general noise are not smoothed away. This is the main advantage compared with applying filters in the spectral domain. It is difficult if not impossible to determine if a deformation in the spectral domain is due to an echo pulse or to absorption lines of a material.

Different types of wavelets can be used for this inverse wavelet shrinkage. We used Daubechies wavelets with 8 vanishing moments, because of their shape being very similar to the shape of the pulses in the time-domain they were particularly suited to distinguish the echo pulses from the rest of the information. The implementation used was Wavelab [36].

Before we continue we want to make a general remark about the topic of computational efficiency and its relevance for this thesis.

Remark 2.3 As done in the previous section, at many points of this work we will

2 Preprocessing

Algorithm 2.2: Proposed Echo Removal with Wavelet Shrinkage

Data: Signal $X_{\text{Total}}(t)$ measured from $t_0, ..., t_n$
Result: Estimation of signal without echoes $\hat{S}(t)$
for $S(t)$ **do**
  Find position of maximal peak $t_{\max}$ $X_1 : [t_0, ..., t_{\max} - 5\text{ps}] \to \mathbb{R}$ with $X_1(t) := S(t)$
  $X_2 : [t_0, ..., t_{\max} + 10\text{ps}] \to \mathbb{R}$ with $X_2(t) := S(t)$
  for $X_i(t)$ *where* $i \in \{1, 2\}$ **do**
    $X := X_i$
    Apply Algorithm 2.1 with level depending hard thresholding
    $\hat{R}_i := \hat{R}$, i.e. the echo pulses of X_i
  $$\hat{S}(t) := \begin{cases} \hat{R}_1(t) & \text{if } t \in \{t_0, ..., t_{\max} - 5\text{ps}\} \\ S(t) & \text{if } t \in \{t_{\max+1} - 5\text{ps}, ..., t_{\max-1} + 10\text{ps}\} \\ \hat{R}_2(t) & \text{if } t \in \{t_{\max} + 10\text{ps}, ..., t_n\} \end{cases}$$
---

propose different methods to solve the same task. Often the more sophisticated method might outperform the simpler one as in the case with wavelet shrinkage and Savitzky-Golay smoothing — if we ignore the artifacts that appear only with wavelet shrinkage. In many cases the less sophisticated method is computationally more efficient though. This is not relevant when we consider only a small set of data that has to be processed. However, in hyperspectral imaging the data amount quickly growth high because with constant width/height ratio to the number of samples grows as the number of pixels in two dimensions. Hence, it grows quickly to a volume where computational efficiency has to be a topic. Therefore, we always propose a second method in comparison that can be used when a high volume of data has to be analyzed.

In the case of windowing after the main pulse instead of the proposed inverse wavelet shrinkage applying a smaller Hamming window can be used to eliminate echo peaks after the main pulse as well. The big drawback is that all information after this echo peak is then lost, the advantage is that a set of possible windows can be calculated beforehand and then simply be multiplied with the data according to the main peak's position. Hence, only one multiplication is necessary per spectrum.

2.3 Frequency-Domain

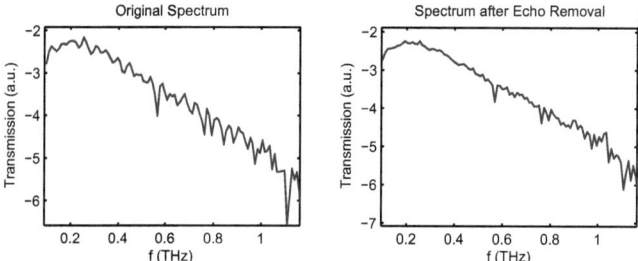

Figure 2.9: Left: Spectrum of 2.7 with interferences of echo pulses; Right: Spectrum of right hand side of 2.8 with removed echo pulses.

2.3 Frequency-Domain

As already described in Section 2.1, standard preprocessing of THz-TDS spectra mainly consists of deconvolving the sample measurement with a reference and calculating the absorption coefficient and the refractive index. It was also mentioned that the absorption index depends on the magnitude spectrum. Therefore, we will describe procedures we apply on the deconvolved magnitude spectrum to gain information about the absorption coefficient where the characteristic absorption lines can be found.

2.3.1 Smoothing the Magnitude Spectrum

In Section 2.2 we talked about interferences appearing in the spectrum that originate from echo pulses in the time-domain. With wavelet shrinkage we achieve a good suppression of these artifacts. The interferences can usually not be eliminated completely and other noise sources can further disturb the measurement quality. Additionally, the sample's thickness influences the SNR and thereby the possibilities to separate the signal from the noise. General noise sources typical for THz-TDS measurements are more concisely considered in Chapter 3. Generally, it can be said that for further data processing the necessity of applying smoothing will arise. Therefore, a suitable method for smoothing the spectrum has to be found.

Depending on the data set the quality of the spectra differs. Especially in hyperspectral imaging, there are spectra which have a very bad SNR — as well

2 Preprocessing

as ones that give clear peaks with smooth dynamic range. Therefore, the goal is not to find an optimal method for a specific application, but rather a method, that works "well enough". This mainly involves two requirements:

(1) Computational efficiency and easy applicability

(2) Preservation of peaks

We will first apply wavelet shrinkage again. It will be used in the spectral domain and with the assumption of Gaussian white noise. Hence, we estimate a global threshold from the first and finest level of wavelet coefficients while in the last section the threshold was chosen especially for each level. As an alternative, we use the Savitzky-Golay filter [40] which is one of the most popular filters in chemometrics. This popularity is founded on its easy and fast calculability and its capacity to preserve peaks [41].

In this method, given a window of points $\{x_{-m}, ..., x_0, ..., x_m\}$ around a center point x_0, one wants to calculate the optimal fit of a polynomial

$$f(j) = \sum_{p=0}^{k} c_{kp} j^p$$

to the given samples. The points must be equally sampled then we can say w.l.o.g. $j = -m, .., 0, .., m$ and the polynomial must be of degree k, where $k < 2m + 1$. Optimality is defined by minimizing the mean squared distance. Hence one wants to find c_{kp} such that

$$\frac{d}{d_{c_{kp}}} [\sum_{-m}^{m} (f(j) - x_j)^2] = 0.$$

Instead of interpolating the polynomial anew in each point, Savitzky and Golay proved that it is possible to calculate the value of the optimal least square polynomial in the central point — i.e. the point to be filtered — by applying a convolution with precalculated coefficients. This is possible because one is not interested in the actual coefficients of the optimal polynomial in each point but just in $f(0)$, i.e. the approximation in x_0. The convolution coefficients then only depend on the size of the filtering window, and the degree of the polynomial. Although it is an old method, it is still used in different areas as an easy to implement and to use tool for data smoothing. There are still alterations being proposed to adapt this method to different applications apart from chemometrics, e.g. geological characterization of spectra [42].

2.3 Frequency-Domain

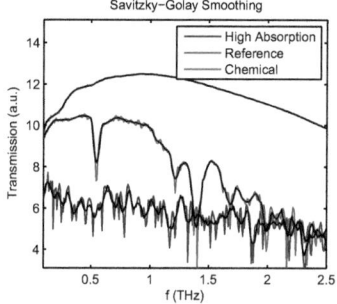 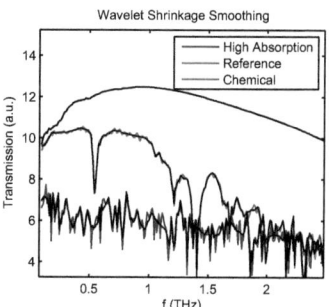

Figure 2.10: Smoothing comparison of two methods on three qualitatively different spectra from one measurement. Black lines are smoothed versions of spectra.

To illustrate the varying quality and SNR of spectra that appear during one data set we show three spectra from one hyperspectral image measurement that are exemplary for this diversity. These different spectra shall visualize the effect of smoothing on different types of spectra. Figure 2.10 presents the two different smoothing approaches that are applied on one very noisy measurement of metallic content, one spectrum containing almost no absorption information (i.e. almost a reference measurement), and one spectrum of a chemical compound. The Savitzky-Golay filter is applied with a broad window size (13 sample points corresponding to ~ 100 Gigahertz (GHz)) and polynomial degree 3 to include extrema as well as inflections. The black curves in the figure show the result of the smoothing. Generally it can be said, that the smoother the spectrum is to begin with the less should be changed. When looking at the reference measurement which is very smooth itself one notes that both filters preserve this spectrum well.

Regarding the chemical compound, both smoothing methods preserve the peaks and filter away small interferences. The wavelet shrinkage, however, preserves the peaks in their whole depth while they are slightly dampened by the Savitzky-Golay filtering. So in this case the wavelet shrinkage performs better. This coincides with the findings of the authors of [43] who did an evaluation of wavelet shrinkage and Savitzky-Golay filtering on different types of peaks and noises. In this paper with artificial noise added, and a known ground truth, the wavelet

27

2 Preprocessing

shrinkage outperformed the other methods as well.

Nevertheless, considering the high absorption measurements we note that in both cases finer structures are filtered out while some of the interferences are still present, but that in the case of wavelet shrinkage the smoothing is done more extensively (consider regions ~ 1.1–1.3 THz and ~ 1.7–1.8 THz) and produces shapes similar to the applied wavelet functions. These shapes are not present in the spectrum beforehand and can therefore be considered to be artifacts.

Although the wavelet shrinkage outperforms the Savitzky-Golay smoothing with respect to peak preservation, we do not choose one of these methods for all application cases but rather decide according to the measurement. The better the measurement and the more controllable the outcome the less the smoothing should change. In case of chemical compound measurements where the exact preservation of peaks is important we will therefore apply wavelet shrinkage. When we measure hyperspectral images with varying content and absorption strength and usually with a higher data volume, we will apply the Savitzky-Golay filter.

2.3.2 Dynamic Range Determination

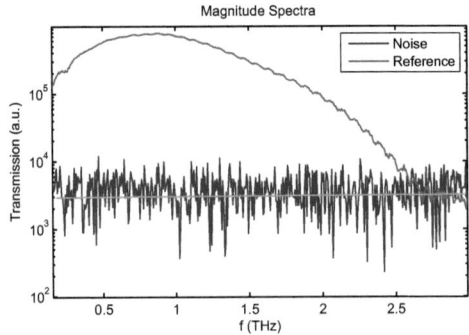

Figure 2.11: Magnitude spectra of reference and noise measurement and calculated noise floor (cyan-colored).

2.3 Frequency-Domain

Once the Fourier spectrum is calculated, we have to determine what part of the spectrum provides useful information and at what point the noise begins and the dynamic range (DR) ends. The most common approach on doing that in THz-TDS is to take a separate noise measurement and thereby determine the noise floor [27]. For this purpose the detector response to a blocked laser beam is acquired. The average of this noise-only measurement is then used as the noise floor [10]. To asses the DR of a measurement all information below this noise floor is filtered. Figure 2.11 shows a magnitude spectrum of a sample measurement in combination with the magnitude spectrum of noise. The cyan-colored line is the calculated mean of the noise measurement depicted in blue. Hence, as the red sample spectrum reaches this cyan line, the rest of the spectrum is considered noise.

However, this method can produce various problems in automatic image processing.

- With respect to spectra of chemical compounds that have absorption peaks deeper than the noise-floor, everything behind this peak will be regarded as being beyond the DR.

- Only the system noise is considered here, other noise sources as for example the samples thickness and interferences appearing for other reasons are not considered. Every measurement must have the exact same conditions as the noise-floor measurement had, otherwise too much or too little information might be cut-off.

- Separate noise measurements are not always taken, so the information is not always given and has to be extracted by other means.

This approach is therefore not satisfying when many different spectra are taken, such as often is the case in hyperspectral imaging applications. Additionally, there is not always the possibility to acquire this noise-only measurement. Nevertheless, even if the noise-floor information is missing it is possible to determine the DR visually. For this purpose mainly two characteristics appearing in all THz-TDS spectra are used:

(1) The interferences have a higher frequency, i.e. of a more random nature, when the noise is approached than when actual content of interest is represented.

2 Preprocessing

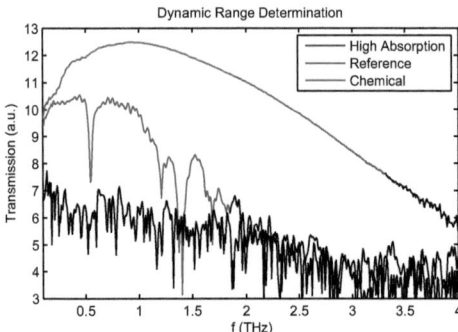

Figure 2.12: Dynamic Range determination by counting extrema after smoothing by wavelet shrinkage. Everything black is assumed to be noise.

(2) The overall shape — or non-constant baseline — is not decreasing as fast anymore but rather staying almost constant.

Our goal was to find a method that relies on these properties and not on a separate measurement. Using the second property, i.e. the almost constant shape of the noise floor has proven difficult and could only serve as an additional indicator for the DR calculations. Therefore, we focused on the frequency characteristics. Again, it seems obvious to use something like the wavelet transform or other frequency based methods. Unfortunately, the frequency characteristic of the peaks is very similar to the noise floor. Furthermore, one drawback of the wavelet transform is that it is not shift invariant, and high coefficients do not necessarily mean high singularities and vice verse. Therefore, the transform is not very robust with respect to varying measurement quality as shown in Figure 2.10.

We therefore propose a simple and effective method that includes information about the application area. Our focus of interest lies on peaks of a specific form, namely the absorption peaks of THz-TDS measurements of solids acquired at room temperature. These peaks have a broad shape of around ~ 100 GHz [44]. For our DR detection we use the fact that when approaching the noise floor and the number of extreme points increases. As soon as we find more than 2 peaks per 100 GHz, we assume that we are facing a noise region. To make the method stable we firstly of all apply previous smoothing if the acquired spectra have a lot of interferences in early regions. Due to the logarithmic shape the noise at

2.3 Frequency-Domain

the end is more pronounced than oscillations in the beginning of the spectrum. Hence, the smoothing methods explained in Section 2.3.1 mainly alter these small oscillations in the beginning. Secondly, the high frequency of maximal points has to be detected over at least 4 consecutive intervals, otherwise we might mistake a very broad, very low peak for a noise region.

Algorithm 2.3: Proposed Determination of End Point of Dynamic Range

Data: Frequency spectrum $Sp : \Omega \to \mathbb{R}$, where $\Omega := \{\omega_0, ..., \omega_{N_{\text{Max}}}\}$ are frequencies in THz. If necessary the spectrum is previously smoothed.
Result: Cut-off point C_{DR} for dynamic range
for Sp *on* Ω **do**
  Interval size: $P_{\text{IV}} := \lfloor \frac{N_{\text{Max}}}{10(\omega_{\text{Max}} - \omega_0)} \rfloor$
  Total number of intervals: $N_{\text{IV}} := \lfloor \frac{N_{\text{Max}}}{P_{\text{IV}}} \rfloor$
  Counter for intervals with more than 1 maximum per 100 GHz: $J := 0$
  for $n = 0$ **to** N_{IV} **do**
    Current interval: $I_n := [\omega_0 + n * P_{\text{IV}}, \omega_0 + (n+1) * P_{\text{IV}})$; Find all maxima Max_n in $Sp(\omega)$ with $\omega \in I_n$
    Count maxima per interval $|\text{Max}_n|$ **if** $|Max_n| \geq 2$ **then**
      if $J \geq 3$ **then**
        Final interval $N_{\text{DR}} := n - 3$
        return Cut-off point: $C_{\text{DR}} := N_{\text{DR}} P_{\text{DR}}$
        break
      Increment Counter: $J = J + 1$
    else
      Reset Counter: $J = 0$

The details of the described procedure are are described in Algorithm 2.3. Furthermore its applicability is illustrated in Figure 2.3 on the spectra introduced in the previous section. The spectra were previously filtered by wavelet shrinkage because we do not have any time-constraints here. The area that belongs to the DR is assigned a color, while noise is assigned black. The high absorption

2 Preprocessing

measurement is considered noise from the very beginning of the spectrum. With respect to the features of interest, i.e. peaks of ~ 100 GHz width, this is correct, however, and this spectrum must then completely be discarded as being too noisy for further analysis.

Discarding a whole spectrum raises the question of how to treat the area after the DR or a pure noise spectrum in automatic data processing. It is necessary to have a common baseline for all spectra that signifies no informational content. Everything after the DR could then automatically be set to this baseline. Producing such a baseline is the topic of the next section. A remark about the shape of the spectra shall be given beforehand.

Remark 2.4 In the previous figures the magnitude spectra of single samples were shown and not the transmittance, i.e. the sample spectrum normalized by a reference as introduced Section 2.1. This was done because smoothing and echo removal is performed separately for sample and reference as well as the dynamic range determination, as this is customary in the literature. For the next section we need the sample to be normalized by the reference, however and from now on refer to the sample spectrum divided by the reference.

2.3.3 Baseline Correction

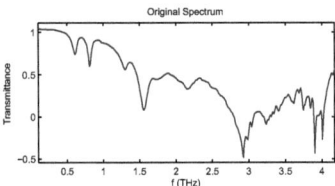

Figure 2.13: Spectrum of p-amino benzoic acid with typical non-constant baseline.

As already mentioned in the previous section, a spectrum should have a constant baseline whenever no informational content is expected. Of course there is informational content in the overall shape of the DR, namely the thickness of

2.3 Frequency-Domain

the material can be determined by that. Nevertheless, this information often is dominant when the difference between two spectra is calculated. Two measurements of the same chemical compound with a different overall absorption would therefore get assigned a big distance. Hence, this overall absorption has to be encoded in another way and the spectrum should be normalized such that the shape of the transmittance is not declining anymore but as constant as possible, while the peaks can be measured from this constant baseline.

A non-constant baseline appears in different kinds of time-series data, therefore there are a variety of methods that are applied for this purpose. Examples for non-constant baselines are infrared spectra, electroencephalograms, and mass spectroscopic data. Approaches on finding such a baseline include empirical mode decomposition (EMD) [45], morphological opening of the spectra [46], or very coarse smoothing or low-pass filtering [47, 48]. There are some characteristics of THz-TDS spectra that make the application of these methods difficult. The problem with EMD is that it is an only data driven decomposition which is not very stable and robust with respect to its decomposition scales. Figure 2.13 shows the transmittance of a chemical compound in the THz domain. One can note that the type and shape of information changes with the frequency and even ends with the DR. The empirical mode decomposition is disturbed easily by such changes and heavily relies on the regularity of extrema. Another important characteristic — again — is the width of the peaks (again see Figure 2.13) which makes the application of morphological baseline correction difficult. Big peaks are smoothed away or at least dampened by the opening and this method is furthermore extremely sensitive to the choice of a structuring element. On spectra with sharper peaks, as for example mass spectroscopic, infrared or magnet resonance data they usually work well.

Because of the described difficulties with state of the art methods for THz baseline correction, we propose a new method for baseline correction that uses a simulated basic shape that is optimally fitted to the spectrum in the least square sense. Additionally it involves a negativity punishment (similar to the negativity punishment for nuclear magnetic resonance data in [48]). In the course of this work Terahertz spectra will be simulated for feature selection evaluation. The particulars of this simulation are explained in the respective Chapter 3. Therein, the first step is to calculate a typical basic THz-TDS magnitude spectrum shape. This shape can be steered mainly by one parameter that is estimated based on the steepness of the main pulse in the time-domain. This coarse approximation of λ_{ini} can be used to initialize such a basic shape.

2 Preprocessing

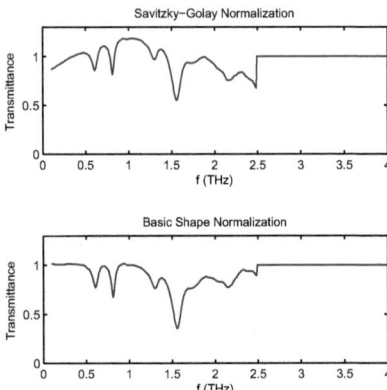

Figure 2.14: Two different baseline correction approaches on the sample spectrum. Top: Savitzky-Golay baseline correction; Bottom: Simulation based baseline correction (Proposed approach) up to DR of ~ 2.5 THz.

Approximating the parameters from the time-domain already gives an idea about the baseline of the spectrum but the estimation is usually still quite far from the correct basic shape. Therefore, an optimization or iteration over various parameters has to be executed. For this purpose we do not optimize over all possible parameters but rather use the first approximation parameter λ_{ini} and iterate over a limited set $\Lambda := \{\lambda_0, ..., \lambda_{\text{ini}}, ..., \lambda_k\}$ around the initial one. The coarseness and range of this set is empirically justified such that the balance between acuteness and efficiency in calculation is met.

Within the optimization we will only iterate over the steepness of the signal. The offset will then be calculated to fit the height of the simulated spectrum to the height of the measured spectrum. We then optimize a distance function between the simulated baseline and the spectrum it should be fitted to. The distance function $D : \mathbb{R}^n \times \mathbb{R}^n \to \mathbb{R}$ that is to be minimized by the optimization

2.3 Frequency-Domain

is defined by:

$$D(X, S) = \sqrt{\sum (S_i - X_i)^2} + \frac{\sqrt{\sum (-|X_i - S_i| + (X_i - S_i))^2}}{3}, \quad (2.6)$$

where $X := \{X_i\}_{i=1}^n$ is the simulation and $S = \{s_i\}_{i=1}^n$ the real measurement of the sample spectrum. In addition to the squared distance between sample and simulation, our distance term involves a penalizer for negativity. It is desirable to have a baseline that is consistently higher than the actual measurement to avoid negative absorption lines.

Finally we propose to combine the baseline correction with the DR determination from the last section. Thus, we can find a baseline for the valid frequency region, i.e. normalize the spectrum in that region and everything that is considered to be noise can be set to the baseline.

Algorithm 2.4: Proposed Calculation of Optimal Baseline

Data: THz-TDS signal S and set of steepness parameters $\Lambda := \{\lambda_i\}_{i=1}^k$
Result: Optimal baseline for logarithmic Fourier magnitude spectrum Sp from signal S with respect to given parameter set Λ and given distance function D

for $n = 1$ to K **do**
  Simulate spectrum with X depending on λ_n
  Set offset-parameter such that $\max(X) = \max(Sp)$
  Calculate $D(X, S)$ as described in Equation 2.6
  if $n == 1$ or $D(X, Sp) < D_{Final}$ **then**
    $D_{Final} = D(X, Sp)$
    $X_{Final} = X$

To show the difference of this approach to a classical method we use a very generously low-pass filtered, i.e. smoothed, spectrum as an alternative for the baseline estimation [47]. For this purpose we apply Savitzky-Golay smoothing. We do not expect inflictions in the baseline which is why a polynomial of degree 2 is used. For the coarseness of the fitting we use a window size of ~ 1 THz.

2 Preprocessing

The top of Figure 2.14 shows the spectrum from Figure 2.13 corrected by the Savitzky-Golay estimated baseline. In comparison to the original transmittance, the spectrum is situated around a target baseline, in this case B_l = 1, but still is far from having a constant baseline. There are peaks higher than 1 and the beginning and end of the spectrum lie far below B_l. In contrast to that our approach for DR estimation and baseline correction on the bottom of Figure 2.14 shows a baseline that is mostly constant, all peaks are below B_l and the spectrum starts from the baseline and almost ends in it as well.

2.4 Conclusion

In this chapter we introduced a number of methods that serve as a toolbox to improve the quality of real-world THz-TDS measurements. Echo pulses in the time-domain can now be filtered away on the one hand. On the other hand, they can be detected, if they are the focus of interest. We adapted two different approaches for echo removal thereby give the applicant the choice between computational efficiency if needed (Windowing) and very good preservation of information besides the peaks (Wavelet approach). An equal situation was presented in the frequency-domain, were we compared two well known smoothing methods. Again the classical method performed faster while the wavelet based method preserved the characteristics of interest best. Furthermore, we developed a method to detect the dynamic range of a measurement that is more adequate for THz-TDS measurements than standard approaches because it includes knowledge about the characteristics of interest and is directly derived from the given data. The latter can also be said about our baseline correction method that is based on specific simulations of the THz-TDS spectra's basic shape. This baseline correction enables comparability in regions were not all spectra have valid information.

3 Simulation of THz-TDS Spectra

In the previous chapter different approaches were explained that can be used to improve the quality of acquired THz-TDS spectra. Nevertheless, the resulting time-series have a dimensionality of several hundred sample points. Hence, in addition to the problem of high data volume, we have to consider the problem of high dimensionality. After the preprocessing from Chapter 2, we assume that the original spectra are representative but contain redundancies. We therefore want to find a representation that preserves the information of interest.

Before we can go into detail with respect to methods that are applied for feature reduction, we need to procure ways how to tell if a feature set is representative. This has to be found out by qualitative but also quantitative evaluation. Data processing methods are usually evaluated only qualitatively or on a very limited number of spectra. We want to add the possibility of a quantitative evaluation. To do so it is necessary to have a sufficiently large data set with known ground truth. One possibility is to create such a data base by measuring different kinds of compounds under different conditions and manually label all these data. It is difficult, however, to be in control of the influencing parameters and to keep the conditions constant during a measuring period. Furthermore it is time consuming and expensive. Therefore, we decided to simulate THz-TDS spectra. Thereby, we are in control of all parameters and always have a ground truth that can be used for method evaluation.

Contribution *In Section 1.1.2 we already mentioned that one of the advantages of THz-TDS is that the excited modes depend on the surrounding media. Unfortunately this can also be seen as a disadvantage because it makes it difficult to build numerical models for the absorption peak positions that fit real-world measurements [49]. In this chapter, a complete simulation scheme for THz-TDS spectra is introduced. The goal is not to simulate spectra of specific compounds but rather simulate spectra that could be spectra of real-world measurements. That means, spectra that incorporate as many of the characteristics of real-world measurements as possible. To do this we divide the simulation scheme into three parts:*

3 Simulation of THz-TDS Spectra

(1) To begin with, the basic shape of the spectra is simulated. There have been various approaches on numerically modeling this basic shape. Two of them shall be presented in Section 3.1.1. In Section 3.1.2 multiple real-world measurements are used to compare the two methods.

(2) In Section 3.2 the kinds of noise that appear in THz-TDS measurements are analyzed and added to the simulated spectra.

(3) The properties of typical peaks are characterized in Section 3.3.1 and the approach on peak insertion is explained in Section 3.3.2.

3.1 Basic Shape

There have been numerical calculations of THz-TDS data. Usually because of the above mentioned difficulties when it comes to simulating the THz response to a certain object, these numerical calculations aim at the basic shape, or rather the undisturbed signal first [50, 51, 52]. These undisturbed signal coincide with what we previously called the reference signal. We therefore start at this point by also first simulating the basic shape, or the undisturbed signal, and later add the additional compounds that are needed.

We will compare two different approaches to forming the basic shape of a THz spectrum. Our goal is to find a basic shape simulation that is simple and at the same time coincides maximally with real-world data. A qualitative comparison will be executed. For that purpose, we will use nine different reference measurements.

3.1.1 Simulation Methods

The first approach is based on [52]. Here, the electric field generated is described as being proportional to the second derivative of a polarization transient, under the assumption of a point source which is the ultra-short laser pulse that is used.

$$E(t) \propto \frac{d^2 P(t)}{dt^2}. \tag{3.1}$$

To simulate a basic form of a spectrum we therefore simulate the transient P. Such a polarization transient in an ideal case is a step function, because the polarization is supposed to switch in an instant. In real-world measurements this

3.1 Basic Shape

is not achievable however. The transient takes some time to happen. Therefore we chose to not use a step function but approximate the transient by a hyperbolic function:

$$P(t) = \tanh(f(t)),$$

with a linear function $f(t)$ and $t \in [-1, 1]$. The steepness of the transient can be controlled by the intercept and slope of $f(t)$. We want to consider the effect of these parameters on the spectra. Figure 3.1 illustrates the simulation of

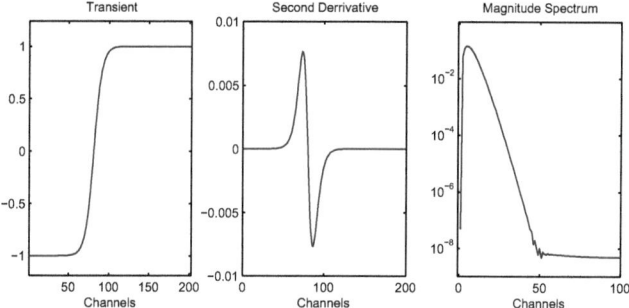

Figure 3.1: Left to right: Simulation of the polarization transient based on Equation 3.1, its second derivative, and its magnitude spectrum.

the transient and the thereafter calculated signal and magnitude spectrum. The parameters effect the height and the width of the measured signal. Physically, the steepness of the transient and the sampling length are determined by them :

$$f(t) = A*t - B, \text{ where } A \leq 1 \text{ and } B \leq A$$

When looking at Figure 3.2 the influence on the magnitude spectra can be seen. While the parameter A influences the width of the spectrum, its height is mainly determined by B.

The second approach we are using is described in [51]. There, the generation of THz-TDS pulses are derived in a physically more detailed manner. We will

3 Simulation of THz-TDS Spectra

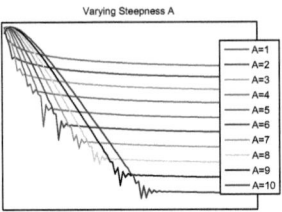
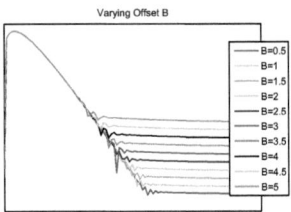

Figure 3.2: Influences of shifting parameters on magnitude spectrum on simulations based on Equation 3.1.

only look on the model from a mathematical point of view though. For a deeper insight into the physical basis of the model we refer to the original paper [51].

The electric field is modeled as the first derivative of the emitter density $j_{\text{em}}(t)$:

$$j_{\text{em}}(t) \propto \exp((C_1\frac{\tau_l}{\tau_e})^2 - \frac{t}{\tau_e}) \, \text{erfc}(C_1\frac{\tau_l}{\tau_e} - C_2\frac{t}{\tau_l}) - \\ \exp(C_1(\tau_l(\frac{1}{\delta_e} + \frac{1}{\tau_e}))^2 - \frac{t}{\delta_e} - \frac{t}{\tau_e}) \, \text{erfc}(C_1\tau_l(\frac{1}{\delta_e} + \frac{1}{\tau_e}) - C_2\frac{t}{\tau_l}) \quad (3.2)$$

with $\text{erfc}(z) := \frac{2}{\sqrt{\pi}} \int_z^\infty \exp(-\tau^2)d\tau$ being the conjugated Gaussian error function. The electric field is then proportional to the derivative of the emitter density:

$$E(t) \propto \frac{dj_{\text{em}}(t)}{dt}. \quad (3.3)$$

Figure 3.3 shows the simulated density with the calculated signal and magnitude spectrum. The parameters influencing the electric field from Equation 3.2 are linked to the following physical characteristics: τ_e is the free carrier recombination time, τ_l the laser pulse duration time, and δ_e the carrier collision time. Figure 3.4 shows that varying the pulse duration time has the strongest effect on the spectrum. The other two parameters only slightly change the outcome of the magnitude spectrum within the region of interest. As only one pulse is simulated, this effect is due to the fact, that τ_l changes the width of the main component. It therefore determines the width of the spectrum, while the other parameters

3.1 Basic Shape

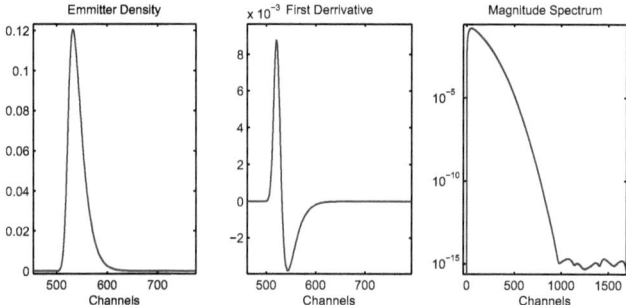

Figure 3.3: Left to right: Simulation of emitter density, based on Equation 3.3, first derivative, and magnitude spectrum.

change the ratio between maximal and minimal peak and the fine shape of the pulse and therefore do not add as dominant information to the spectrum. The parameters in Figure 3.4 are chosen according to typical measuring data.

3.1.2 Evaluation

On the one hand, the latter method has a more detailed physical derivation, therefore choosing the parameters in accordance with a special measurement is easier. On the other hand, the simulation itself takes longer. We will now compare the simulations with real-world data of reference measurements. These measurements do not have any characteristic peaks and therefore should contain only the basic shape information.

The initial set of parameters, even if cleanly physically derived, usually does not fit to the real-world measurements perfectly [51]. Hence we use the physically derived parameters only as starting points for an iterative approximation of a good parameter set. We do not look for the optimized data set, but want to get a general idea which of the basic shape simulation schemes fits the kind of data we deal with. In addition, we want to determine which simulation scheme to use for further research.

In Figure 3.5 the Euclidean distance between simulated basic shapes and original spectra is visualized. On the right hand side one can see the basis for this

41

3 Simulation of THz-TDS Spectra

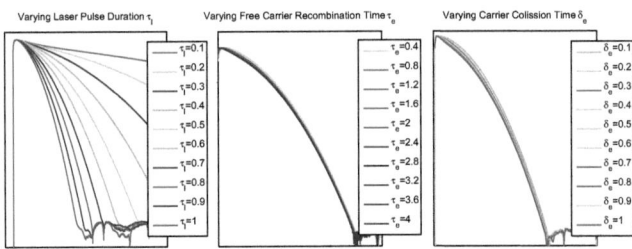

Figure 3.4: Influences of shifting parameters on magnitude spectrum of simulations based on Equation 3.3.

Duvillaret	7.9	3.8	7.1	10.3	5.9	4.6	3.8	3.8	4.6
Hyperbolic	7.7	2.3	4.5	11.4	6.7	2.4	2.3	5.1	2.4
Difference	-0.2	-1.5	-2.6	1.1	0.8	-2.2	-1.5	1.3	-2.2

Table 3.1: Euclidean distances between two proposed simulations (with optimal parameters) and nine original spectra. Bottom: Difference between top two distances for each reference measurement, smaller zero means hyperbolic approximates better and vice verse.

distance calculation on the example of one measured spectrum and the simulations with the best parameters. The best parameters are chosen by calculating the best fit as described in the baseline correction approximation in Section 2.3.3. The first impression is that the hyperbolic function simulation produces results closer to the original spectra. We use a t-Test to test the hypothesis that the expectation value μ_H of the distance from a hyperbolic function simulation is smaller than the one of the Duvillaret function simulation μ_D, i.e. that the difference is smaller than zero: $\mu_H - \mu_D \leq 0$. In Table 3.1 the distances and their respective differences are plotted. The mean $\overline{D}$ is -0.7778 and the sample's standard deviation s_D is 1.5442. Therefore, $t = \sqrt{9}\frac{-0.7778}{1.5442} = -1.511$ and with a confidence of 90% the hypothesis that μ_H is smaller than μ_D can be accepted.

It has to be said, that in [51] a comparison to measured data is executed as well. Some adjustments are performed due to the flaws of the simulation. Including these adjustments leads to a better result than the original numerical model.

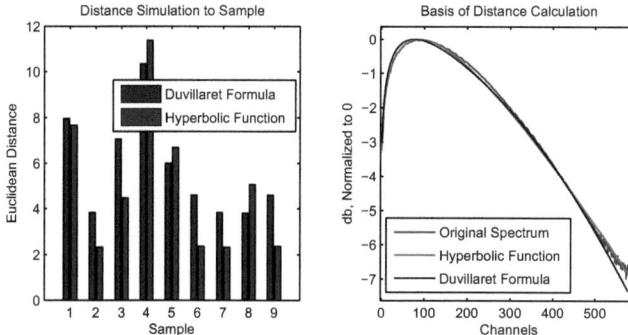

Figure 3.5: Qualitative comparison of two simulation techniques for basic shape of a THz-TDS spectrum.

Nevertheless, incorporating these further steps adds to the numerical cost of the whole procedure. One point of our simulation is though, that to be able to simulate a big volume of data, the calculation time can not be neglected altogether. The density from Equation 3.2 that is used for Equation 3.3 involves considerably more operations than Equation 3.1. Hence, with our test data, finding the appropriate parameters in Equation 3.3 needs over 250 seconds while Equation 3.1 has found a good parameter set in less than one second. For the cause of data simulation the quality of both approaches is sufficient and we therefore do not apply further refinement steps to improve 3.3 but use Equation 3.1 for its better performance and better numerical efficiency.

3.2 Noise

Now we have to include typical noise that can appear in THz-TDS measurements into the simulation. We will analyze what noise appears and why it appear. However, the noise that is considered here will be exclusively measurement noise. No environmental noise will be included (for example water vapor will not be a topic, see Remark 2.2) as its variation can be too big to be within the scope of this work.

In [53] the authors showed that in THz-TDS the variance of the transmission

3 Simulation of THz-TDS Spectra

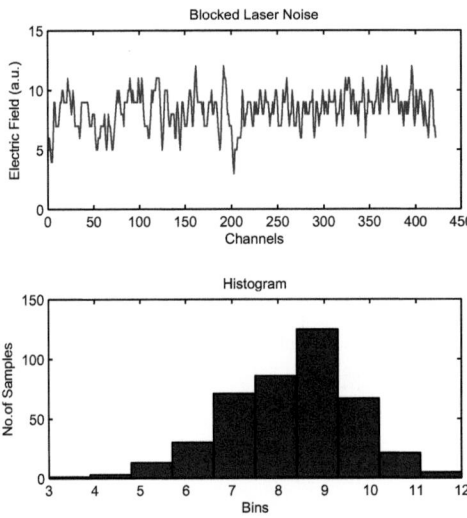

Figure 3.6: Top: Distribution of blocked laser signal; Bottom: Histogram of values.

modulus can be expressed depending on emitter, detector and shot/laser noise. The laser is a strong noise source in THz measurements and enters the THz setup through the antennas on the emitter and detector side. On the emitter side it is the dominant noise source during a THz pulse ([53], there called "emitter noise"), however, we do not explicitly treat it in our study because it is present only during a very short period of time. With sufficiently long waveforms as measured for spectroscopy, detector noise is increasingly important. It is also partly generated from laser noise but has other components particularly from electronic noise in the antenna and preamplifier. The relative intensity of these noise sources depends on the laser model, the antenna type, the current-to-voltage resistor in the first amplifier, and the modulator ("chopper") frequency.

To determine the characteristics of this noise we have to take measurements

3.2 Noise

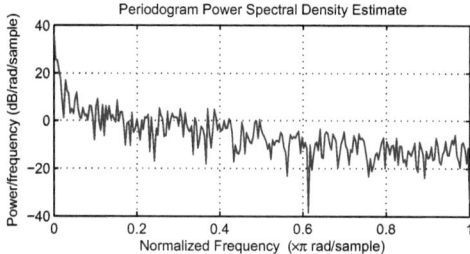

Figure 3.7: Power spectral density of blocked laser beam noise. Its $\frac{1}{f}$ character can be seen.

with a blocked laser beam. These are not taken separately but are taken directly from the signal. The component before the laser pulse of a reference measurement is used for this purpose. These measurements are taken without a sample between emitter and detector. Figure 3.6 shows the noise signal and its histogram. The shape suggests that the noise is normally distributed, with its mean and standard deviation as parameters. We use a χ^2-Test to verify this hypothesis [54]. The χ^2-value here is 17.07 for the blocked laser noise histogram. With an α-level of 99% and 7 degrees of freedom, this is within the acceptance region that is limited by 18.48, hence we can conclude that the blocked laser beam noise is normally distributed.

In addition to the distribution we analyze the power spectral density. In Figure 3.7 it can be seen that the density is not constant over the spectrum but declining with increasing frequency. The respective noise form is called $\frac{1}{f}$-noise [55].

This suggests that the laser is the dominant noise source and therefore, the noise that shall be used in our simulation is $\frac{1}{f}$ noise with a Gaussian distribution.

We now add the noise to the simulated signals. The simulation is done according to the model described in [56]. The dynamic range of THz measurements is determined by the reference measurement normalized with the noise floor and typically moves in a range of 10^3 [57]. We simulate the noise with respect to that and the result can be seen in Figure 3.8.

45

3 Simulation of THz-TDS Spectra

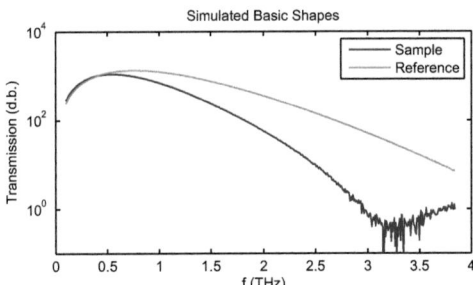

Figure 3.8: Logarithmic transmission amplitude of reference and sample's basic shape with $\frac{1}{f}$ noise added.

3.3 Peaks

Absorption peaks in spectra of molecules appear because of the transition between different modes. THz radiation can excite intermolecular bond vibrations, large mass molecular oscillations and phonon, i.e. collective vibrations [20]. Due to this, crystalline structures can be identified which is particularly useful in drug detection and security control (see Section 1.1.2). Most pharmaceutical coating material, for example, is semi-transparent in contrast to the agents within. Thereby, monitoring of medication production is possible [58]. For a deeper insight into the genesis of molecular modes and the resulting absorption lines we refer the reader to [59] but for the purpose of simulation mainly information about the position and shape of the peaks is needed and shall be given now.

3.3.1 Shape of Peaks

The first thing one can observe when considering THz-TDS spectra such as the p-amino benzoic acid on the left hand side in Figure 3.9, is that the peaks or absorption lines are not very sharp. The reason for this is that the considered materials are solids. In solid material one usually observes interactions of many atoms at the same time. That leads to a superimposition of their absorption bands, i.e. to their collective width [60]. On the one hand it is possible to sharpen these lines by cooling the sample. In [44] the authors achieved a sharpening of some features down to 15 GHz by a special preparation of the sample in combination with cooling it down to 11 K. On the other hand, most THz-TDS spectra are measured at room temperature, because the most prominent features remain visible and the technology is relatively insensitive to thermal interferences in contrast to other technologies. This is one of the main advantages over conventional infrared spectroscopy [24], for example. Therefore, we will consider the type of peaks that appear at room temperature measurements. The width of these peaks is described as ranging around 100 GHz and this is the width that will be used here.

Regarding the position, depending on the compound, the characteristics can be found in the whole THz spectrum. The dynamic range of our simulated measurements — as well as most real-world THz-TDS measurements — is set in the regions between ~ 100 GHz up to ~ 3 THz. Therefore, we will simulate peaks in that region. Additionally, many compounds of interest, such as explosives and drugs can be identified via their properties in that area [61].

The number of characteristics is varying greatly as well. Some compounds have only one absorption peak while others have several. This also depends on the sharpness of the peaks. In [44] with the sharpness of the detected peaks also the number increased from 1 up to 18 for a specific compound. At room temperature due to the low resolution the number of characteristics remains under ten.

The depth of the peaks depends on the concentration of the agent on the one hand [24] and again the temperature [44] on the other hand. The lower the temperature and the higher the concentration, the deeper the peaks. Even though we assume room temperature, i.e. a constant temperature, the depth is still influenced by the concentration. It can range from complete absorption to only minimal absorption.

3.3.2 Peak Insertion

The peak position and depth can be chosen arbitrarily between 100 GHz and 3 THz and 5% and 100% absorption. We want the simulated peak to be as smoothly integrated into the basis shape as possible and not to have any ringing effects on their edges. Cubic splines are a fast and simple way to achieve these goals. For each peak position 10 of the simulated sample points 50 GHz left and 50 GHz right of the peak position are used as supporting points, as well as the peak position point multiplied by the chosen absorption relative to the basis spectrum. In formal terms: If $X = \{x_1, ..., x_n\}$ is the simulated spectrum with

3 Simulation of THz-TDS Spectra

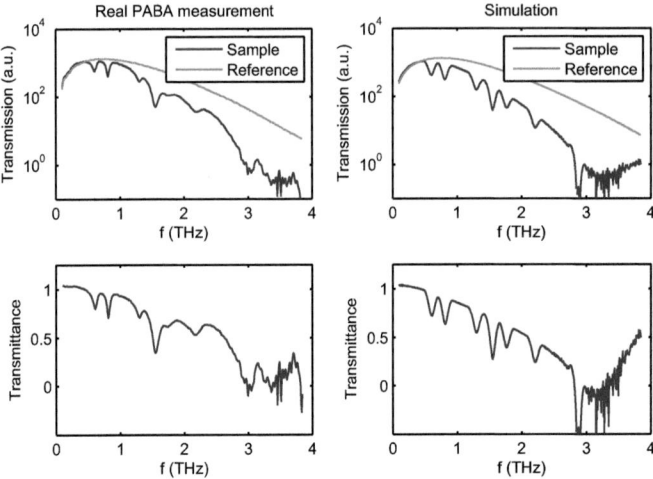

Figure 3.9: Left: Measured Spectra; Right: Simulated spectra; Top: Magnitude spectra of reference and sample; Bottom: Transmittance (sample spectrum divided by reference).

the frequency channels x_1 to x_n ranging from 100 GHz up to 3 THz, and x_{Peak} is the peak's center, and x_{left} and x_{right} the channels 50 GHz left and right of it respectively, then $\{x_{\text{left}-10}, ..., x_{\text{left}}, x_{\text{Peak}} * \text{Absorption}, x_{\text{right}}, ..., x_{\text{right}+10}\}$ are the supporting points used for the spline interpolation.

The result of the simulation can now be seen in Figure 3.9. To be able to qualitatively compare a simulated spectrum with a real one, the left hand side of the figure shows a measured spectrum of a pressed p-amino benzoic acid pellet and its reference spectrum. On the right hand side a simulated spectrum with the same number and position of peaks is shown. Below that, the respective transmittance spectra are plotted. This should serve as an illustration of the similarity of the simulated spectra with measured ones. The peak simulation is done using splines.

3.4 Conclusion

With the proposed simulation scheme we can generate a testbase that can be used to evaluate data processing methods. We are able to produce spectra with a specific signal to noise ratio and the most dominant kind of noise as well as a desired number of peaks with whatever depth and position wanted.

In the next chapter, concepts for THz-TDS feature sets that have a lower dimensionality than the full spectra will be introduced. With the simulation proposed here, we will be able to systematically validate these methods by simulating THz-TDS spectra and by comparing a reduced feature set with the now known ground truth.

3 Simulation of THz-TDS Spectra

4 Feature Reduction

The goal of this chapter is to find an appropriately representing but dimensionality reduced feature set for THz-TDS data. In Chapter 2, the data was preprocessed in such a way that time-series belonging to similar compounds have a similar spectral outcome regardless of interferences such as echo pulses or different overall absorption and dynamic range. The dynamic range determination, for example, additionally reduces the dimensionality. Nevertheless, the remaining feature dimension is still high which has various drawbacks:

- Computational cost of data processing:
  The high dimensionality influences the calculation of distances between objects (see Section 5.3.1) and the possibilities to apply classical image processing for spatial denoising. Regarding the former, information from all channels needs to be used to add to the distance information. In the latter, the spatial denoising needs to be applied to all channels or even three-dimensionally which results in a cubic increase in computational effort.

- Biased result of the distance calculation:
  When applying the introduced distances all the channels are equally weighted, as there generally is no prior information as to where information of interest lies. Undesired effects may appear when the informational content of the channels differs greatly. When using Minkowski norms, for example, small distances in a high number of channels lead to similar effects as big distances in only a few channels, although these phenomena should be considered differently.

- So-called "curse of dimensionality" [62]:
  With growing dimensionality the distribution of sample points in space becomes more sparse.

- Over-fitting to irrelevant features [63]:
  Information that might be of no interest in a specific content can dominate the important information.

4 Feature Reduction

For these reasons, extracting relevant features shall be the topic of this chapter.

Contribution In the preliminary section the data model that builds the basis of the further considerations will be explained. As the identifying characteristics can be found there, the main focus lies on the representation of the magnitude spectrum. After discussing the different possibilities of feature reduction, we analyze the principal component analysis (PCA) as one of the state of the art methods in Section 4.2.1. Because of the drawbacks of existing approaches, the main contribution of this chapter is the introduction of a wavelet based feature reduction in Section 4.2.2. The principal advantage of said feature set is that it does not only preserve discriminating information but also the position of this discriminating information. Furthermore, it can be applied incrementally. We expand this approach by proposing to use complex as well as real wavelets in Section 4.2.3.

The simulations of the previous chapter are utilized to evaluate the feature sets. First the optimal basis function and parameters for the real wavelets are analyzed in Sections 4.3.1. In Section 4.3.2 the two approaches are compared with the full spectrum by calculating the correlation of all three feature sets with the known ground truth.

In addition to the magnitude spectrum, other features should be included. The most important ones are discussed, among them are representations of the phase in Section 4.4.2. All proposed and standard features are resumed at the end of the chapter in Excursus 2.

4.1 State of The Art

We will first give necessary definitions about the data model and then subsume the most important methods that are used to reduce the dimensionality of time-series data. There are two main approaches to reduce the dimensionality of time-series data that can be roughly categorized in feature extraction and feature selection. In feature selection the reduced feature set is a subset of the original features while in feature extraction the original features are transformed into a lower dimensional space or a space in which the content can be represented in a more simple way.

Many papers in the THz domain mainly focus on data acquisition. Therefore, the standard methods explained in 2.1 together with some simple features such as the main amplitudes height and offset in the time-domain [64, 65] are often the only feature extraction that is done. Although transforming the spectra to the Fourier domain is already a form of feature extraction, this approach principally preserves the number of features. Hence, the number of frequency channels in the Fourier domain is still high. If the content of interest is known beforehand, choosing a reduced set of frequency bands is possible [66, 67], but it would be desirable to find a representation that can be used to also classify unknown content. For this purpose up to date mostly the whole spectrum is used or the sum of values of equally sized intervals [61].

We therefore aim at finding a method that represents the whole spectrum on the one hand and on the other hand reduces the dimensionality drastically.

4.1.1 Data Model

Before we give a short review of methods used in other time-series application areas, we will give some definitions about the data model that will further be used.

Although often use analogously, objects do not coincide with the features that describe them. Matters can be described in many different ways and the current description, while hopefully representative, is usually not comprehensive and often can be enlarged as well as reduced or normalized. However, we will start by assuming an entity to be described by a finite number of different features from one dimensional feature spaces. This multidimensional representation shall be called the pattern of the entity [68]. We will now define mathematically what a representation of an entity is.

Definition 4.1 Let O be a finite subset of the set of possible objects $\mathbb{O}$. We call the mapping $X := a(O)$ with $a : \mathbb{O} \to \Omega$ the q-dimensional **representation of** O where $\Omega := \Omega_1 \times \Omega_2 \times ... \times \Omega_q$ is the q-dimensional feature space consisting of one dimensional feature spaces Ω_i. Accordingly, each entity $o \in O$ is said to be represented by its pattern $x := a(o) = (\omega_1, ..., \omega_q)$.

Remark 4.2 We assume that all objects that are to be compared are represented in Ω, i.e. are comparable in all features.

Each of the one-dimensional feature spaces Ω_i can be nominal, ordinal or numerical. But only numerical data will be considered here, as the objects in this thesis are represented by spectra or signals, i.e. time-series of some form. Therefore, it is assumed that $\Omega_i \subseteq \mathbb{R}$, thus, the sample set $X := a(O)$ is a subset of $\mathbb{R}^q$.

4 Feature Reduction

Accordingly a set of n patterns $X := \{x_i\}_{i=1}^{n}$ corresponds to a $q \times n$ matrix, with the rows describing all the features and the columns describing the patterns or samples. This matrix is called the sample matrix.

4.1.2 Feature Selection

In feature selection, all existing channels have to be reviewed in some way to find the ones that best represent the data. The evaluation criterion that measures the optimality of the respective subset varies. Generally, one can discriminate between so-called filter and wrapper methods. Filter methods evaluate the feature subset by calculating some residual between the selected and the full set while wrapper methods combine the choice of features with the supervised or unsupervised classification approach. Filter methods are generally faster. Nevertheless, there have been efficient applications of wrapper methods in high data volume areas as well [69, 70]. Especially if training data is present, wrapper methods combine feature selection with the supervised classification. This approach has lead to good results also in hyperspectral imaging applications [71]. However, in combination with unsupervised classification, wrapper methods and filter methods do not differ much, as they generally use some entropy criterion to evaluate a channel subset. In one case this would be an entropy measure between classes and in the other case an entropy measure between features themselves [72]. A typical filter method is the so-called "best subset selection", an efficient way to evaluate all features by a specific criterion. Unfortunately, due to its computational costs it is generally only applicable on less than 40 dimensional spaces. Alternatively similar solutions can be retrieved faster by backward propagation or forward selection [73].

The evaluation criterion in spectral imaging is mostly related to peak characteristics, as it is the case in popular approaches such as peak picking or line fitting. Those are often used in mass spectroscopy due to the sharpness of the peaks [74]. In this approach the peaks of interest have to be determined first. Then all samples or spectra are compared by their values at these peak positions. The characteristics of the peaks, such as left- and right-hand gradient, or the peak value in comparison to the average, amongst others are used to determine the quality of the chosen subset in this feature selection method.

However, this method does usually not work well for THz-TDS spectra of solids. This is due to the fact that the peaks that appear in this region are broad and merge with one another. In contrast to that, in mass spectroscopy they are

4.1 State of The Art

sharp and singular. In THz-TDS, however, it is difficult to find criteria for peak determination that give stable results. There are algorithms where peak picking is used to identify various peaks in the time-domain [75], or in gas-state or extremely cooled measurements, since the peaks are sharper when the sample is cooled or in gas-state. But, as the possibility to acquire at room temperature is one of the most important advantages of THz-TDS we will focus on room temperature measurements of solids. Hence, peak picking is not a practical option for our applications.

Subsuming the previous paragraph, we will not use feature selection as a tool for feature reduction in THz-TDS of solids for various reasons:

- Computational costs and poor scalability

- Increase of features in incremental data acquisition, if characteristics can appear in all original channels

- Instability of classical methods.

4.1.3 Feature Extraction

In feature extraction the original feature space is transformed in such a way that most of the information is represented in a different feature space. This transformation is carried out either just to improve the visibility of features interest — such as the transformation from the signal to the Fourier domain applied on THz-TDS data does [27] — or to additionally reduce the number of dimensions. In time-series data different kinds of transforms are used: The Fourier transform, the cosine or the wavelet transform as well as polynomial representations. Next to such a transform based approach other techniques are based on eigenvalue decompositions such as the singular value decomposition or the principal component analysis [76, 73].

Feature extraction techniques in general are a more promising advance for the THz-TDS spectra than feature selection, due to their non constant shape and the somewhat blurry features and the flaws of feature selection. We will therefore now analyze the possibilities of applying such methods on THz-TDS spectra.

4 Feature Reduction

4.2 Representing the Magnitude Spectrum

It has already been mentioned that the content of interest, namely the characteristic absorption peaks can be found in the frequency-domain. Therefore, only little information is extracted directly from the time-domain and the frequency-domain is mostly used. For that purpose the Fourier transform is applied. The resulting complex-valued information is analyzed separately by the refractive index (representing mainly phase-information) and the absorption coefficient (representing magnitude information). The refraction spectrum is important for dielectric properties and the analysis of meta-materials and semiconductors. However, in the analysis of chemical compounds, the main information lies within the absorption — represented by the magnitude — as described in Chapter 2.

The transformation of the time-domain channels to the frequency-domain is represented by a transformation of the initial feature space Ω^T to a new feature space $\Omega^F \subset \mathbb{R}^{q_F}$. During further preprocessing, the dynamic range calculation in particular cuts down the number of usable channels (to the maximal dynamic range of all spectra in the object set O). Nevertheless, the remaining dimensionality $q^{F*} \leq q^F$ is still very high. It usually ranges around ~ 400 spectral channels.

To reduce the dimensionality we want to evaluate feature extraction possibilities. We will start with showing the application of an eigenvalue based method and illustrating its flaws. Thereafter we will propose and evaluate a new transform based method.

In eigenvalue based feature extraction, the data set is analyzed with respect to its entropy. Information with high entropy will be preserved while information with low entropy can be discarded. One of the most popular algorithms in this area is the principal component analysis (PCA). In order to demonstrate that methods such as PCA are not fully applicable on the THz-TDS data, we will analyze this method in more detail further and illustrate its functionality on our application spectra.

4.2.1 Feature Extraction by Principal Component Analysis

In PCA the information of interest is assumed to lie within the channels that have the highest variance. The high variant information is found in the following way: With a singular value decomposition a centered $q \times n$–matrix X can be decomposed to a diagonal $n \times n$–matrix D and the orthogonal $q \times n$–matrix U

4.2 Representing the Magnitude Spectrum

and the likewise orthogonal $n \times n$–matrix V:

$$X = UDV^t, \qquad (4.1)$$

where the entries of D are ordered decreasingly such that $d_1 \geq d_2 \geq \ldots \geq d_n \geq 0$. The sample covariance matrix is given by $\frac{X^t X}{q}$ within which, due to Equation 4.1, $X^t X$ can be expressed as

$$X^T X = VD^2 V^t. \qquad (4.2)$$

Derived from this decomposition, the variables $z_j := Xv_j$, with $j = \{1, \ldots, n\}$, are called the principal components of X. They are linear combinations of the original features with decreasing variance. From Equation 4.2 one can derive that

$$\mathrm{Var}(z_1) = \mathrm{Var}(Xv_1) = \frac{d_1^2}{q}.$$

Hence, we have $\mathrm{Var}(z_1) \geq \mathrm{Var}(z_2) \geq \ldots \geq \mathrm{Var}(z_n)$.

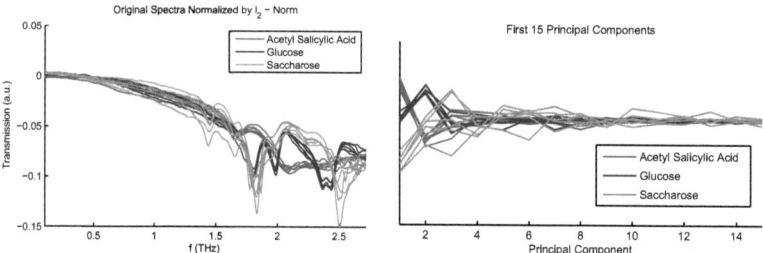

Figure 4.1: Left: Measurements of three different compounds; Right: First 15 principal components (PCs) of the spectra on the left.

With this method, a decomposition is achieved that shifts the high-variant information into the first dimensions while the latter dimensions only contribute little new information.

PCA can be applied to reduce the dimensionality of THz spectra. On the left hand side of Figure 4.1 example spectra of three different compounds are

4 Feature Reduction

shown, namely acetyl salicylic acid (ASS), glucose, and saccharose. We use a hyperspectral measurement consisting of 36 signals per compound but for reasons of interpretability only 6 of each are plotted in the images. To obtain a good basis for the PCA the dynamic range is cut-off generously and the spectra are normalized with the l_2 norm. The resulting first 15 Principal Components (PCs) are shown on the right hand side of the same Figure 4.1.

Transforming high variant information into the first components works well on the displayed spectra. The first PC already separates the three kinds of chemicals from one another. The later channels contain progressively less information. However, the PC decomposition of the saccharose measurement displays a higher variance compared to the PC of the glucose and the ASS. As the PCA only searches for high variant information, this influences the decomposition quality altogether. This problem of instability becomes more dominant when the spectra that are used as a basis for the decomposition are not well prepared, i.e. contain many noisy channels. On the left hand side of Figure 4.2 one can see the same spectra from Figure 4.1 but with a less sharp cut at the valid frequency region. The respective PCs are shown in the same figure on the right hand side. Here, discrimination between the different chemicals is far more difficult than it was in Figure 4.1.

This demonstrates that noisy channels frequently add so much variance information without containing discriminating content that the variance alone cannot be used as an indicator for the different compounds. The instability of PCA

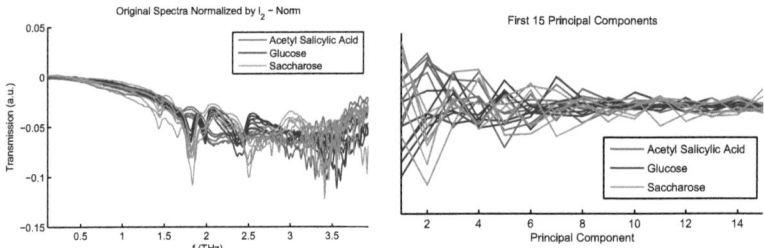

Figure 4.2: Left: The three measurements from the left hand side of Figure 4.1 with a wider dynamic range. Right: Respective PCs for these noisier spectra.

4.2 Representing the Magnitude Spectrum

is particularly disrupting in hyperspectral imaging. In this area the amount of materials is high and diverse, materials with different overall absorption, a low dynamic range, and a high noise level might appear.

In addition to this instability there are other reasons why the PCA may be an unfavorable choice for feature reduction in THz-TDS:

- The conventional PCA works only on closed data sets. An already calculated transformation cannot be used for newly acquired data. New data may necessitate a complete recalculation. Even though there are various incremental approaches on estimating the PCA but most of them have an unsecure approximation error [77].

- With the calculation of the principal components, information on the original location is lost. Although one might correctly separate two spectra, the information where the differences lie, i.e. what lead to this separation, is not evident from the transformed components.

- There are efficient methods to calculate eigenvalues and therefore the PCA. Nevertheless, the calculation still has to be carried out on the full data set. It is not possible to perform it incrementally. Therefore, with increasing data volume the covariance matrix grows quadratically with the sample size and hence all calculations do. The algorithm is hardly parallelizable and does not scale well. If many imaging measurements have to be compared a complete analysis might be impossible due to computational limitations, as described in the context of hyperspectral infrared imaging [20].

- A variance based approach is not adequate for data with a high variation in quality. Furthermore, problems caused by the necessary normalization can arise. Normalizing with the l_2 norm leads to a suppression of the undesired variance effects described in Chapter 5 Section 5.3.1 (l_2 in combination with a Euclidean distance calculation leads to the same distance results as the described cosine similarity). At the same time, however, normalization levels out differences in absorbing characteristics, i.e. , deep peaks will be damped, low peaks will be sharpened.

Due to the several issues that can arise when using PCA the necessity of choosing a method that focuses on the shape of the characteristics of interest rather than the general variance seems advisable. Furthermore, for high data volume the

4 Feature Reduction

method should scale well to be applicable on imaging measurements. Finding such a reduction approach will be the topic of the next sections.

4.2.2 Feature Extraction by Using Real Wavelets

So far, we have applied the wavelet transform at different points in this thesis, for echo pulse detection in the time-domain as well as for peak preserving smoothing in the frequency-domain. There are reasons to assume that the wavelet transform is also adequate for the representation of THz-TDS samples. due to their similarity to THz characteristics, the easy calculability of the transform and the reconstruction properties, wavelets have been used for signal representation in the time- as well as in the frequency-domain [78]. Most applications focus on denoising, however. In contrast, we propose to use them for feature extraction in the frequency-domain.

Wavelet coefficients for feature reduction purposes have been utilized before in time-series applications such as [79] where Daubechies wavelets are used to show possibilities of sparse representation of THz images in the time-domain. In [80] certain wavelet coefficients are chosen by swarm intelligence methods. In [81] coefficients from a so-called competitive wavelet network are chosen with a neural map. The drawback these approaches have in common is that they do always work on a closed data set. Well operating features of one data set may not be applicable on another. Either one has to recalculate or increase the number of features again.

We therefore propose a different way to choose wavelet coefficients. In [82] the authors use all Haar wavelet coefficients of one specific scaling level to represent information from time-series. Depending on the time-series application, the respective scaling level can be fixed. This is done by considering the size of characteristics that have to be detected. As mentioned above, the size of the peaks in THz-TDS spectra of solids is very broad in relation to the width of the spectrum, e. g. about 100 GHz within a 100 Ghz to 3 THz spectrum. It should therefore be possible to use only very broad frequent information and still be able to detect those peaks. By choosing one downsampling level, information of the respective frequency will be depicted. The filter model is determined by the choice of wavelet basis function, the mother wavelet. As explained in Chapter 2 Remark 1, the wavelet transform separates frequency and time-information by representing the original signal in different scaling levels. The higher the level, the finer the information. Thus, with a signal of $2^J = N$ channels there are J scal-

4.2 Representing the Magnitude Spectrum

ing levels with the $(J-1)$st level containing 2^{J-1} coefficients that represent the high-frequent information, while the last two coefficients of level 0 contain only $2^0 = 1$ coefficient with low-frequent information. The coefficients are retrieved recursively by a high-pass filtering on the basis of the mother wavelets.

Therefore, the main idea of the proposed feature reduction is to choose an adequate wavelet basis function and represent the information contained in the spectrum by wavelet coefficients of a low downsampling level. As the peak characteristics in THz-TDS spectra have a low frequency, the characteristics should be detected in the 5th or even the 4th level. This reduces the dimensionality to 32 or 16 coefficients.

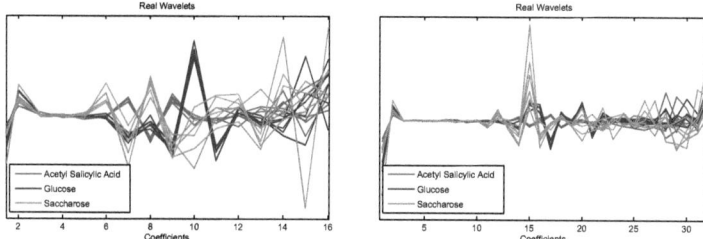

Figure 4.3: Wavelet coefficients of 4th (left) and 5th (right) downsampling level on the basis of Daubechies 8 wavelets from the spectra shown in Figure 4.2 (left).

In Figure 4.3 the 4th and 5th level of the wavelet decomposition of the example spectra in Figure 4.2 are shown. In contrast to the PCA the wavelet transform is not as sensitive to noise in the later regions of the spectra. The discriminating information around 2 THz can be identified while the similarities around 1 THz are represented as well. Thereby the dimensionality is reduced from previous 512 to 32 or 16 channels, additionally the location of features or singularities is represented. Due to the shape of the basis function, these singularities can be detected, and additionally the information is normalized to zero. To illustrate the robustness, Figure 4.4 shows the spectra without previous normalization. Normalization is necessary to yield a good PCA-result. In contrast to that, Figure

4 Feature Reduction

4.4 shows that normalization has only little effect on the wavelet coefficients, represented here by the 4th level coefficients. This means that for the application of the proposed real-wavelet based feature reduction neither baseline correction nor normalization are necessary. This can be especially useful in the case of hyperspectral imaging where the variety of compounds is potentially high. Hence, the baseline correction could fail, due to too much noise, and the normalization could suppress information of interest.

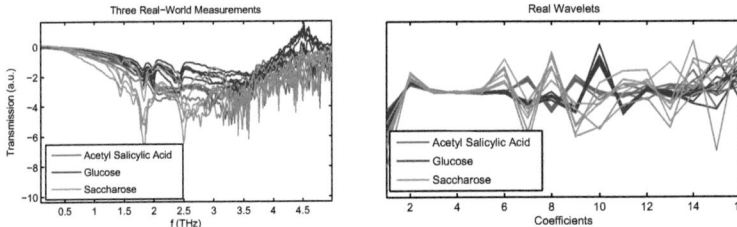

Figure 4.4: Left: Spectra from Figure 4.2 (left) without previous l_2 normalization; Right: Their wavelet coefficients from level 4.

Additional advantages of the wavelet based feature selection are:

- Scalability. The feature sets are calculated separately for each spectrum, the method is therefore highly parallelizable and scales linearly with the number of features. Because of the non-existing dependencies very high-volume data can be processed.

- The separate calculation of the features leads to comparability of incrementally processed measurements. No recalculation is necessary.

- The positions of the differences as well as similarities can be identified.

- No normalization or baseline correction is necessary. The wavelet decomposition detects singularities of any kind. If the baseline is non-constant but behaves smoothly this will be within the low-pass information of the wavelet decomposition. Thereby, the singularity information and the overall

4.2 Representing the Magnitude Spectrum

absorption are separated. This is an advantage in comparison with working with baseline corrected information because the information about the height of the magnitude — i.e. the overall absorption — is not lost, but rather stored in the coefficients of the remaining scaling factor.

- There is a variety of wavelet basis functions that can be used as the basis-filter. By appropriately choosing this basis, the sensitivity towards noise is diminished and only features that resemble the peak shape of interest are represented.

- The wavelet transform itself is easily implemented and calculated.

We have now seen that there are many advantages of using real wavelets for THz-TDS feature reduction. However, the real wavelet transform has some well known drawbacks as well. Although generally the real wavelet transform is a valid representation to identify differences, it can produce various problems. We will focus on two specific ones.

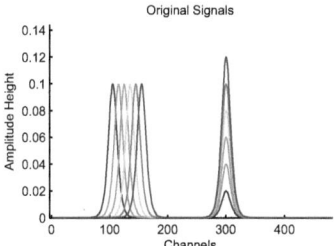
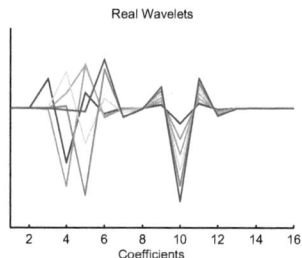

Figure 4.5: Left: Simulated peaks with offset and different height, Right: Corresponding real wavelet coefficients of level 4.

(1) Wavelets are bandpass filters therefore they oscillate around singularities. Hence, even if the singularity is detected and represented by coefficients that are different from non-singularity areas, the actual height of the coefficient does not directly reflect the height of the singularity.

4 Feature Reduction

(2) The wavelet coefficients are not shift invariant, i.e. the same singularity will usually produce different coefficients when shifted in the original domain.

To illustrate these two effects we simulated a number of peaks. The simulation can be seen on the left hand side of Figure 4.5. On the left of this plot there are peaks that have the same height and shape and are shifted slightly. On the right we simulated peaks that have the same position but different heights. The right hand plot of Figure 4.5 shows the wavelet coefficients of level 4 (on the basis of Daubechies 8 wavelets) of these time-series. The coefficients of the shifted peaks are partly very different. The dark red and the bright red peak, for example, although only slightly shifted, have very different outcome, one showing no oscillation, the other a high degree of oscillation. The peaks with the same position and different height are represented in a continuous way, but clearly oscillate, i.e. they have a positive as well as a negative part, although the peak only has one direction.

4.2.3 Using Complex Wavelets

The above mentioned problems with the real wavelet transform are well known and different approaches on overcoming them have been developed. The stationary wavelet transform, for example, provides a shift invariant version. It has the great disadvantage, however, to contain as many coefficients in each level as original samples. It is therefore computationally expensive as well as not dimensionality reducing and hence not suitable for our purpose [33]. Other approaches that have a special focus on the development of a complex version of the real wavelet transform were developed in the late 90s. In a complex transform one would want the shift and the height information to be separated as it is the case with the Fourier transform where phase and magnitude represent qualitatively different information. While the phase gives information about shifts in certain frequency regions, the magnitude indicates the height of the respective frequency.

One computationally efficient approach in that area was developed in 1998 [83], it is called the Dual-Tree Complex Wavelet Transform (DT CWT). Choosing some coefficients from such a DT CWT to represent a closed data set best has been applied in hyperspectral imaging previously [84]. But again analogously to the real wavelet based feature set, we focus on choosing a fixed feature set for all THz-TDS spectra. To calculate a complex wavelet transform one can execute two real wavelet transforms with a special set of filters each [83]. One

4.2 Representing the Magnitude Spectrum

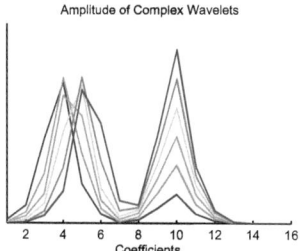
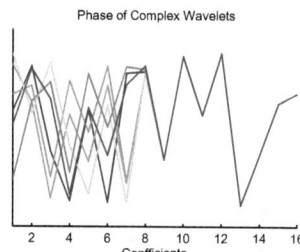

Figure 4.6: Left: Magnitude spectrum of DT CWT of simulated peaks of Figure 4.5, Right: Phase spectrum. Note that the left shows the height of peaks, while the right shows the shift between peaks.

of these transforms provides the imaginary part and the other one the real part of the coefficients. The two filter sets are based on wavelets ψ_h and ψ_g that are approximately analytic in their complex form $\psi_h + i\psi_g$ and one is approximately the Hilbert transform of the other. With such filters, the forward transform as well as the inverse transform is easily executed using only real arithmetic. The inverse transform consists in switching the real and the imaginary part and using the same filter bank for backward transformation [85].

In Figure 4.6 the magnitude and phase of the DT CWT are shown. The magnitude spectrum is an indicator of the height of the peaks. First of all there are no oscillations in the wavelet domain. Second, it is hardly influenced by the shifts in the original domain, only in adequate shifts in the magnitude. In contrast, the phase represents only the shift information. That means, the phase completely differs where the peaks are shifted, i.e. in the beginning and where the peak position is the same and only the height differs, the phase shows no differences at all (as in the later peaks).

Hence, with the complex wavelet transform we can describe the THz-TDS data by separating information about peak position and peak intensity. However, the transform has the drawback of producing twice the amount of coefficients to represent the spectra as the real wavelet does. One option to reduce the dimensionality again would be to just use one of the indicators that can be calculated, for example to only the magnitude or only the phase. As these two parts express such different content, we rather combine them. When calculating the distance between two spectra on the basis of such a representation it is sensible to use a

4 Feature Reduction

function that somehow combines both information. A natural choice for such a combination is to use the intersection between these two indicators, i.e. use the smaller distance of the two between each two samples. Choosing the minimal of two distances to build another one is a classical example of a T-norm if some other prerequisites are satisfied.

Definition 4.3 A mapping $T : [0,1] \times [0,1] \to [0,1]$ is called **T-norm** if it satisfies

- $T(x,y) = T(y,x)$, i.e. commutativity
- $T(x,T(y,z)) = T(T(x,y),z)$, i.e. associativity
- $T(x,1) = x$ and $T(x,0) = 0$
- If $x < y$ and $v < w$ then $T(x,v) < T(y,w)$, i.e. monotonicity

To use T-norms to combine information from different distances leads to the following assumptions: If two things are the same with respect to one (i.e. have distance 0) then the combined distance assumes them to be equal as well. When we apply this function we always use the smaller indicator as the guide value. On the other hand when both values are decreasing the joint one cannot increase which complies with the general demands one would have to such a distance function combination.

The function $T(a,b) := min(a,b)$ with $a,b \in [0,1]$ is the most classical T-norm and we will use it to combine the distances calculated on the basis of the different coefficient sets. As a and b are the results of the distance calculation between a phase and a magnitude spectrum they do not naturally take values in $[0,1]$. Hence, a normalization is necessary. For this work we only consider the two different distance measures we use here, namely the Euclidean distance and the cosine distance. The cosine distance is easy to normalize, as it takes only values in the interval $[0,2]$ (see Equation 5.3). A division by 2 restricts the distance to $[0,1]$ and the T-norm can be applied.

Using the Euclidean distance makes normalization slightly more difficult, a and b are still bigger than 0 but generally not smaller than 1. The phase spectra can only take values in the interval $[-\pi, \pi]$, hence the distance between two phase spectra is limited by a maximal distance in each channel of 2π and the total number of channels. Therefore, depending on the chosen decomposition level j, dividing the distance by $\sqrt{2^j(2\pi)^2}$ restricts the result to $[0,1]$. The distance

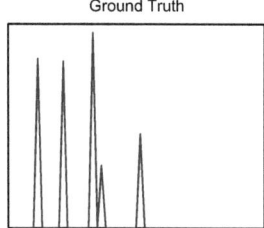 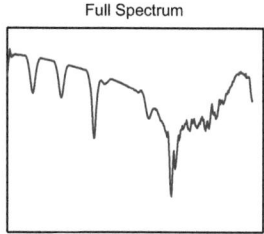

Figure 4.7: Schematic view of the different shapes that are being compared.

between the amplitude spectra on the other hand is not bounded, for practicality reasons we therefore normalize the distances of a sample set by its maximum. With the thus defined T-norm we have a distance measure that combines the phase and the magnitude information of two spectra.

4.3 Correlation Based Evaluation

In this section we will analyze if the feature sets that were proposed in the previous section represent the spectra accurately. In Chapter 3 we have proposed a simulation scheme where the basic shapes of the spectra are simulated and the peaks are inserted in the frequency-domain. The ground truth that has to be the basis of any evaluation scheme, is therein represented by idealized spectra consisting exclusively of the peak positions and their absorption intensities.

In Figure 4.7 the left hand side shows this ground truth of a 5 peak spectrum consisting of the position and depth of the inserted peaks. There is no noise and no declining baseline in this idealized spectrum. The right hand side shows the respective simulated magnitude spectrum with the same peak positions and depth but with the typical shape and noise of THz-TDS spectra.

For the feature set evaluation one has to compare every proposed set with the ground truth (i.e. with the left hand side of Figure 4.7 rather than the simulated spectra). This comparison is done with respect to the distance between two spectra under the different representations, i.e. the wavelet representation, the full spectrum, and the ground truth.

4 Feature Reduction

We simulate repetitively 25 spectra of 5 different "compounds", i.e. 5 different peak feature sets that are applied to in 5 different realizations of the basis spectrum each. The comparison is done by analyzing how much the distance induced by the given feature set is correlated with the distance induced by the ground truth. The distance between the ideal or ground truth (GT) features is denoted by D_{GT}. The distance matrix between the feature sets (FS) is denoted by D_{FS}, both of which are symmetrical $n \times n$ matrices. For the correlation analysis Pearson's correlation coefficient [86] is calculated between each two columns $D_{\text{FS}}(\cdot, j)$ and $D_{\text{GT}_{(\cdot,j)}}$:

$$C(D_{\text{GT}_j}, D_{\text{FS}_j}) = \frac{\sum_{i=1}^{n}(D_{\text{GT}_{(i,j)}} - \overline{D_{\text{GT}_{(\cdot,j)}}})(D_{\text{FS}_{(i,j)}} - \overline{D_{\text{FS}_{(\cdot,j)}}})}{(n-1)(\text{std}(D_{\text{GT}_{(\cdot,j)}})\text{std}(D_{\text{FS}_{(\cdot,j)}}))}$$

where $\text{std}(D_{\text{GT}_{(\cdot,j)}})$ and $\text{std}(D_{\text{FS}_{(\cdot,j)}})$ are the standard deviations of the jth column of the matrix D_{GT} and D_{FS} respectively. Furthermore n is the number of samples, i.e. 25 simulated spectra. The accuracy measure is then defined as the mean correlation over all columns $\overline{C(D_{\text{GT}_j}, D_{\text{FS}_j})}$.

In the further course of this section this correlation calculation will be used to evaluate all the proposed feature sets.

The real wavelet transform has the advantage of being adaptable to different applications by the variety of wavelet basis functions. However this adaptability also leads to the necessity to choose among these basis functions. Therefore, before analyzing the performance of real wavelets in comparison to the other two feature sets, namely complex wavelet transform and full spectrum, we have to determine which parameters to use optimally. For that purpose we will devote the next sub-section to finding the best basis-function and downsampling level for our application.

4.3.1 Best Parameters for Real Wavelet Decomposition

First of all we want to know which of the two downsampling levels provides the best results. In Section 4.2.2 we demonstrates that the full coefficients of the valid frequency region of the 4th and 5th downsampling level should represent the 100 GHz peak information. Both levels are surveyed.

Figure 4.8 shows the results of this survey, i.e. the correlation analysis of the 4th level (~ 16) and the 5th level (~ 32). For this figure, we simulated 3-peak spectra with varying absorption on the basis of Symmlets with different

4.3 Correlation Based Evaluation

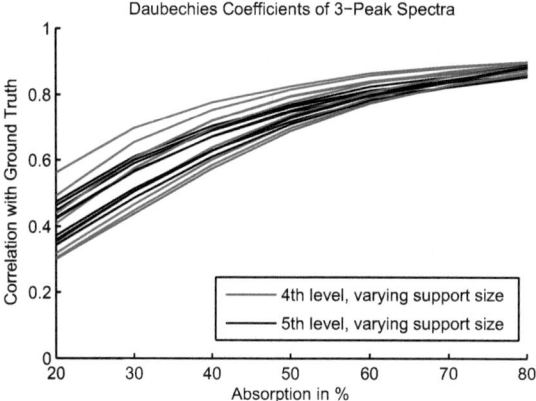

Figure 4.8: Correlation of 3-peak spectra with varying absorption depth for 4th and 5th level coefficients of Symmlet-decomposed spectra with varying support size.

support sizes. Exemplary parameters are chosen. The results of simulations with a different number of peaks or another basis function are qualitatively similar with respect to the downsampling level.

The figure illustrates that the use of the 4th level coefficients, i.e. fewer coefficients, consistently yields better results. With alternating support sizes we can achieve a better correlation with the coarser 4th level coefficients than with the finer 5th level ones. Only in the very high absorption areas, the results are similar, but still the best performing 4th level features give better correlation with the ground truth. In addition to better performance in this analysis, the computational efficiency is higher with fewer coefficients. Therefore, there is no reason to regard the 5th level coefficients as superior to the 4th level ones and we will from now on use the 4th level coefficients for further investigation.

In a next step we will analyze which wavelet basis function performs best in this correlation analysis. For all methods applied in this thesis we use the Wavelet package [36], therefore, we analyze the different mother wavelets that are provided by this library. The basis functions under consideration are orthogonal wavelets with a compact support, such as Daubechies wavelets, Coiflets, and Symmlets

4 Feature Reduction

as well as symmetrical wavelets with an infinite support such as Battle-Lemarie wavelets. For more detailed information on the characteristics of these wavelets we refer to the vast literature on this topic such as [87].

First, the optimal support size for each wavelet basis function is chosen in order to be able to compare the different basis functions with each other. For the analysis we again fix the number of peaks per spectrum and calculate the correlation for different absorptions.

In Figure 4.9 we show the simulation results of a correlation analysis of the different support sizes of Symmlet basis function, Coiflets, Daubechies wavelets, and Battle-Lemarie wavelets. First of all the basis function as well as the support

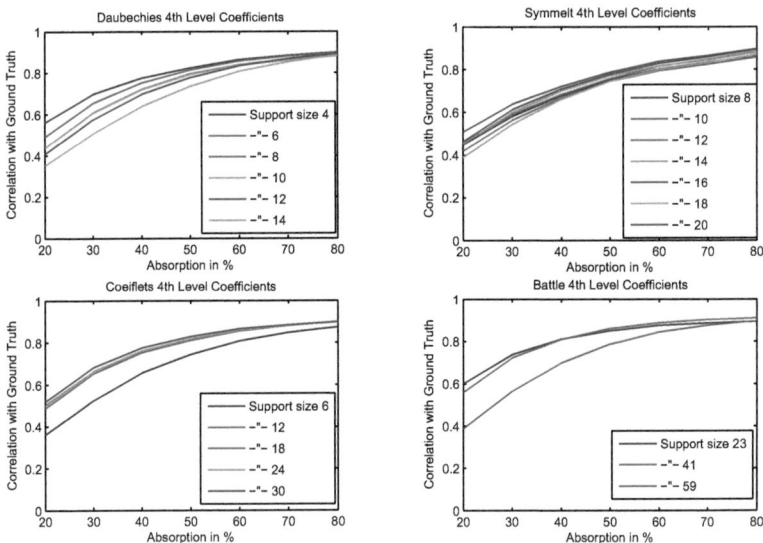

Figure 4.9: Comparing dependency of correlation of 4th level wavelet coefficients on support size of respective basis function.

size have a significant influence on the correlation outcome. Neither a higher nor

4.3 Correlation Based Evaluation

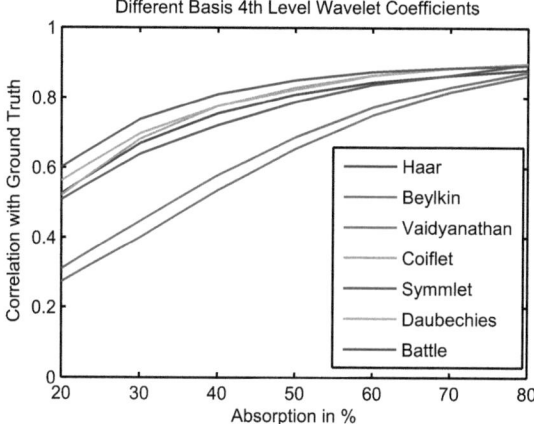

Figure 4.10: Correlation of 3-peak spectra 4th level wavelet coefficients based on different wavelet basis function.

lower support generally leads to better results. The appropriate choice depends on the respective basis function. For Daubechies wavelets the smallest support size of 4 gives best results, for Symmlets a relatively high one of 16, for Coiflets even the highest one of 30 and for Battle-Lemarie wavelets, which after all have a compact support due to discretization, a support size of 23 is optimal.

We now use these basis functions with their best parameters to compare them with each other as well as with non-parametrizable basis functions.

The result of this comparison of different wavelet basis functions can be seen in Figure 4.10. Vaidyanathan and Beylkin wavelets perform worst. Hence, they should not be chosen for a feature reduction of THz-TDS spectra. The other basis functions all perform similar. However, the Battle-Lemarie wavelet performs better compared to the others and will therefore be used for all the feature reduction based on real wavelets in this thesis.

We have illustrated in this section that there is a considerable variety of wavelet basis functions that can be chosen to model THz data. We also have shown the importance of the specific basis function and of the support size for the quality of the feature extraction. One should be carefully consider the different options

4 Feature Reduction

when choosing wavelets for feature reduction. For the complex wavelets, different basis functions can theoretically be chosen. In the implementation we used [83] we found that the choice of basis function hardly had any influence on the correlation analysis. Therefore, we used the default parameters for the further comparison.

4.3.2 The Three Feature Sets

We will now compare the three feature sets which are:

(1) The full spectra with about 350 channels

(2) The 4th level of the real wavelet based decomposition with the Battle-Lemarie basis function with support size 23

(3) The minimum T-norm of the phase and the magnitude of the 4th level complex wavelet decomposition

The correlation analysis is carried out by considering the dynamic range of the simulation, i.e. about 350 channels. Therefore, to represent the 4th level coefficients we do not use the full $2^4 = 16$ calculated coefficients but the equivalent to the dynamic range of the full spectrum. The basis of the wavelet transform are 512 channels, i.e. the representation of the dynamic range is done by $\frac{16}{512} * 350 = \sim 11$ channels. Therefore, in the case of the real wavelet representation we use only 11 channels instead of 350, and 22 for the complex wavelet transform.

Figure 4.11 displays the results of this analysis. In total, 10 000 spectra were simulated. The normalization that is necessary for the T-norm to combine phase and magnitude information from the complex coefficients was done by normalizing the (Euclidean) distance matrix of each simulation set of 1000 spectra by its maximal and minimal distance.

Both reduced feature sets perform better than the full spectrum does. Hence, in addition to it being computationally more efficient to use the reduced feature sets, the peaks are better represented. Furthermore, as expected, clearer peaks, i.e. peaks with a high absorption can be more easily detected by all feature sets than peaks with very low absorption. However, the advantage of the wavelet coefficients in comparison to the full spectrum correlates positively with decreasing absorption of the peaks. In the lower regions the real wavelet coefficients achieve a correlation of 60% while the full spectrum ranges below 30%.

4.3 Correlation Based Evaluation

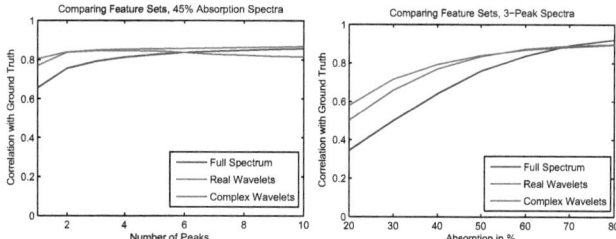

Figure 4.11: Correlation analysis of feature sets, number of features reduced from 350 to 11 (real) and 22 (complex wavelets). Left: Fixed absorption, varying number of peaks; Right: Fixed number of peaks, varying absorption.

Although all feature sets give satisfying correlation values, reducing the feature set with either method, i.e. complex or real wavelet transform, is beneficiary in comparison to the full spectra.

Taking into account only the results displayed in Figure 4.11, we now can suggest to use real wavelets for the representation of THz-TDS spectra. They show the best correlation with the ground truth and provide a feature reduction from 350 to 11 features while the information of time as well as location is preserved. Nevertheless, the complex wavelet transform should not be discarded completely. It outperformed the full spectrum and also provided a good dimensionality reduction. Furthermore, using phase and magnitude combined by a T-norm might not even be the most advantageous choice under these circumstances. Only the separate consideration of the two spectral compounds brings out the benefits of the complex transform, which is the more direct interpretation in comparison with real wavelets. It may lead to better results, for example if specific information about peak height or peak shifts is necessary. One would then combine the magnitude and phase information in a weighted way.

73

4 Feature Reduction

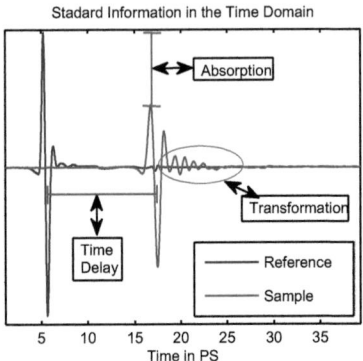

Figure 4.12: Relative height of main amplitude (absorption) and set-off (time-delay) are the first features used for visualization.

4.4 Additional Signal-Features

With the wavelet based feature sets we have found an appropriate representation of the magnitude spectrum of THz-TDS spectra. But, although the most important information for the identification of chemical compounds measured in solid state lies within the magnitude spectrum, it is sensible to account for information from the time-domain as well as the phase.

4.4.1 Time-Domain Features

If a THz-TDS imaging measurement is acquired, one needs an easy visualization that shows if the measurement works in principal. For this purpose, features are used that show the outline of some of the imaging content already and are easily calculated directly from the time-domain. For this purpose the relative height of the main amplitude are frequently used as well as its set-off.

In Figure 4.12 these two features are shown. To calculate them, a reference measurement is needed in addition to the sample. The ratio between the main amplitude height and the time set-off in picoseconds are used.

A further feature from the time-domain that can be used, is the position of echo pulses. Echo pulses are useful to characterize the thickness of semi-translucent

4.4 Additional Signal-Features

materials since they are mainly caused by reflections within the material. The isolation of the echoes from the signal deformation by wavelet shrinkage as proposed in Section 2.2.2 can be used for frequency-domain smoothing on the one hand. On the other hand it can be used for echo detection to be able to use the echo positions as features.

4.4.2 The Phase Spectrum

One of the advantages in THz-TDS is, that the Fourier transform of the time-domain signal provides us with the phase as well as the magnitude spectrum. There are many applications where mainly the phase spectrum is used. Especially when characterizing dielectric materials such as isolating layers on semiconductors, phase information is highly relevance. In [88], for example, the authors use THz phase information for an efficient characterization of such materials.

However, when characterizing chemical compounds the identifying peaks are primarily found in the magnitude spectrum and little phase information should be present. Generally the phase of a THz-TDS measurement is at least piece-wise linear, due to the single-cycle pulse measured [28]. The linearity is usually interrupted at the points where absorption lines lie. This effect can be seen in Figure 2.3 in Chapter 2 on the right hand side. Hence, it is possible to extract peak information from these shifts in the phase. However, the information about the depth of these peaks is less stable in this domain than in the magnitude. Therefore, usually the latter is used to extract this information. As a representation of phase information that cannot be found in the magnitude we propose to use an approximation of this linear shape. For each unwrapped phase spectrum we find the straight line that fits the spectrum best in a least square sense. Its slope and its intercept are then used as two further features.

After this short section about the time-domain information that belongs to standard representations of THz-TDS spectra we will now outline the possibilities to include spatial information into the analysis of hyperspectral images.

---

2 Excursus (The Proposed Feature Set (PFS))
We will frequently refer to all of the features we explained and introduced in the previous sections, therefore we will recapitulate these features in a list and call them the Proposed Feature Set (PFS) and from now on may refer to these PFS features by their number as well as their name.

4 Feature Reduction

The first two features are extracted from the time-domain

(1) The maximal amplitude height of the time-domain

(2) The maximal amplitude offset in the time-domain with respect to the reference

As representatives for the phase of the Fourier spectrum of measurements of solid state chemicals we use

(3) The phase slope

(4) The phase intercept

To represent possible reflections in the measurement we use

(5) Position of first echo pulse, relative to the position of the main pulse

Representing the basic shape of the magnitude of the Fourier spectrum we use

(6) The last scaling factor of the wavelet decomposition, either a real wavelet decomposition or a complex one.

The main part of this chapter was concerned with representing the magnitude of the Fourier spectrum well. We found two kinds of representations that can be used. Both are based on some kind of dyadic downsampling scheme. Therefore the exact number of features varies depending on how well the dynamic range coincides with a power of 2 number of channels.

We can either use:

(7–.) Real wavelet coefficients of 4th level, in total equal or less than 16. Battle wavelets and Daubechies wavelets with the smallest respective support sizes yielded the best results in the feature evaluation scheme.

or alternatively use

(7–.) Complex wavelet coefficients of 4th level, in total quay or less than 2 × 16. The complex wavelet coefficients are split into phase and magnitude, therefore we have double the amount of coefficients in comparison with the real wavelets

4.5 Conclusion

In this chapter we have introduced different features that can be used to represent THz-TDS measurements. We particularly proposed a representation of the magnitude spectra that holds various advantages over classical approaches such as the principal component analysis. This representation consists of the wavelet decomposition of the spectra at a certain scale. We applied the real wavelet transform as well as the complex wavelet transform. While the real wavelet transform can be adapted by choosing between different parameters, the complex wavelet transform can supply feature information that can be easily interpreted.

An extensive simulation based evaluation of the real wavelet based feature reduction was performed. We found out that using a subset of the 4th level wavelet coefficients yielded the best results of all level based subsets. Various wavelet basis functions performed well. Among the best were the Battle-Lemarie wavelets, the Daubechies wavelets and Haar wavelets. For the further analysis both Battle-Lemarie as well as Daubechies wavelets are used.

The simulation was additionally used to evaluate the complex feature set. For the evaluation purpose we combined the information from the two coefficient sets given by magnitude and phase with a T-norm. The correlation analysis in comparison with the real wavelets and the full spectra resulted in the two proposed wavelet based feature sets being superior to the full spectral data. The real wavelets with optimally chosen wavelet basis function and window size performed best. Nevertheless in applications with non-ideal measurement conditions the application of only the complex magnitude will be advantageous for data interpretation as the shift invariance and directionality of the complex transform can be easily interpreted.

Further features from are proposed to be used to identify material by their refraction or reflections.

4 Feature Reduction

5 Automatic Segmentation of Spectral Data

If the data volume is too large for manual organization, automatic methods are required that correctly detect similarities and subsequently organize data into segments or classes. For this purpose classification is used. Although classification in general has been a topic ever since computational data processing exists, it still remains a field of active research, and there are still new algorithms and methods being developed. This is due to the growing amount of data available in digital form as well as the growing variety of types of data.

When the data volume is high, it is sensible to split the classification process into two parts. In Figure 1.3 in Chapter 1 we already showed the separation into segmentation — or unsupervised classification or clustering — and classification. Segmentation is used to group the spectra into classes containing similar items — according to some similarity measure based on their features, i.e. according to inherent characteristics. Supervised classification then is used to assign fixed classes to the object according to some exterior criterion, such as "being an explosive" in the case of security control, or "having a defect" in case of quality inspection. As the latter strongly depends on the specific application, we will consider only the former, i.e. clustering, in this thesis. Classification can then be applied on the found segments that are regions of interest.

Unsupervised classification is an approach to find such a segmentation as it is a technique that relies only on the unlabeled data and its features. A distance measure has to be chosen and classes or clusters are then build by minimizing the intra-cluster distance and maximizing the inter-cluster distance.

Contribution *In Section 5.1, an overview of clustering algorithms will be presented and the focus on hierarchical methods motivated. The necessary definitions for the clustering model that enable us to describe the basic algorithm are then given in Section 5.2. The basic hierarchical clustering algorithm will be explained as well as how different sets of distances influence it. Furthermore, in Section*

5 Automatic Segmentation of Spectral Data

5.3.1 we propose to use the cosine distance for THz-TDS spectra instead of more established ones for that area.

The problem of high data volume will be the topic of the next section. We justify the use of the Chameleon algorithm in Section 5.4. We then analyze the algorithm further and propose alterations to the initial similarity measure in Section 5.4.2. In Section 5.5 we suggest an altogether alternative clustering approach that uses the basis of Chameleon with the classical hierarchical clustering.

To improve the segmentation quality of hyperspectral images we will discuss the inclusion of spatial features in Section 5.6. Two different approaches will be used. In Section 5.6.1 we discuss methods that apply classical image processing on single slices and suggest an algorithm based on this idea while in Section 5.6.2 we propose to include a spatial distance directly into the clustering algorithm.

In the last part of this chapter we will discuss methods that can be used to evaluate a clustering when a ground truth is given. As the existing approaches are somewhat unsatisfying for methods that do not assume manual interference, we propose an alternative evaluation scheme in Section 5.7.3 and a feasible implementation for this scheme in Section 5.7.4.

5.1 State of the Art

How can we organize a given set of time-series objects in an automatic manner and in such a way that in the end, objects that are similar to each other are in the same group while objects that are different from each other are in separate groups. This is the object of cluster analysis. The literature provides a substantial variety of clustering method and still algorithms are being developed. This is due to the abundance of objects that can be characterized in many different ways. Furthermore, the resulting categories can vary greatly in shape and representation, depending on the application. The existing algorithms are typically divided into two groups, hierarchical clustering (HC) and partitional clustering (PC). We will first outline the characteristics of PC, its advantages and also its flaws. Thereby, the use of HC in our work will be justified.

In PC the goal is to distribute patterns to different classes. Probably the most popular PC method is the k-means algorithm. Here, starting with K initial centers, all data points form clusters by joining their nearest center point. On the basis of the thus formed clusters, centroids are calculated which become the new centers. The process is repeated until the center points do not move anymore.

Although in worst case scenario k-means has exponential performance [89], in average case it has a complexity of $O(NK)$ and gives good results even in high volume data sets [90]. K-Means is one of the most frequently used clustering techniques. Nevertheless, its main drawback is that the number K of clusters as well as initial cluster centers have to be set beforehand. Moreover, the clustering result strongly depends on how they are initialized. Many other partitional methods, even ones that are quite different in their characteristics such as artificial neural networks [91, 90], share this initialization problem [70, 92].

Avoiding the parameter choice problem is the main reason why we will focus on HC. Additionally the outcome of an HC algorithm is not just a partition of the data, but a whole hierarchy of partitions. This hierarchy contains rich information about the process of grouping the data. The actual partition that is needed for further processing can then be chosen in accordance with the partition hierarchy. This is usually easier than setting parameters beforehand. A disadvantages of HC is the computational expense of these approaches.

5.2 Clustering Model

We will now give the necessary definitions that enable us to correctly describe HC. Before we talk about algorithms themselves, we have to define on what basis these algorithms should work. Therefore, we will start by describing various terms, such as partition and hierarchy, mathematically. While the definition of a partition of a data set is a standard definition in discrete mathematics, hierarchies are usually not defined that clearly. The topic of clustering is often only considered from the computational side, therefore the used terms may sometimes be ambiguous. For this reason we will now give definitions for all the basic terms the reader needs, to understand the algorithms properties. The definitions are accumulated from different sources, mostly [93, 68, 94]).

Definition 5.1 Given an finite set O, a set of subsets $\mathscr{P} := \{C_i\}_{i \in M} \subseteq \wp(O)$ is called a **partition of** O if and only if

(1) $C_i \cap C_j = \emptyset$, for all $i, j \in \{1, ..., m\} := M \subset \mathbb{N}$ if $i \neq j$

(2) $\bigcup_{i \in M} C_i = O$

Partitions can be compared to each other by defining a *finer than* relation. This is necessary to build hierarchies that are the result of a HC.

5 Automatic Segmentation of Spectral Data

Definition 5.2 Let $\mathscr{P}_i$ and $\mathscr{P}_j$ be partitions of the set O. Then $\mathscr{P}_i$ is said to be **finer** — " $\vartriangleleft$ " — **than** $\mathscr{P}_j$, i.e. $\mathscr{P}_i \vartriangleleft \mathscr{P}_j$ if and only if for all $x \in \mathscr{P}_i$, exists $y \in \mathscr{P}_j : x \subseteq y$.

Remark 5.3 Often the subsets $C_i \subseteq O$ are assumed to be non-empty. Under this condition, the set of all partitions together with the *finer than* relation build so-called lattice, which has some desirable algebraic properties. Many clustering algorithms produce empty clusters, however and to account for this fact we have to neglect this condition. Nevertheless the set of all partitions of O will be denoted by $\mathfrak{P}_O$, it is a subset of $\wp(\wp(O))$.

In contrast to other unsupervised learning techniques HC does not require the number of classes to be known beforehand. This is due to the fact that the result of such an algorithm is not just one partition of the data, but rather a sequence of such partitions, a so-called hierarchy.

Definition 5.4 Given a set of partitions $\{\mathscr{P}_i\}_{i=1}^h \subset \mathfrak{P}_O$ with $h \in \mathbb{N}$. Then $\mathscr{H} := \{\mathscr{P}_i\}_{i=1}^h$ is called a **hierarchical partition or hierarchy** if $(\mathscr{H}, \vartriangleleft)$ is totally ordered. W.l.o.g. we assume the labeling to be ordered, i.e. $\mathscr{P}_1 \vartriangleleft \mathscr{P}_2 \vartriangleleft ... \vartriangleleft \mathscr{P}_h$.

Many HC algorithms additionally produce an index containing information about the heterogeneity of the partition:

Definition 5.5 The **clustering index** is a function $f_h : \mathfrak{P}_O \to \mathbb{R}^+$, which holds for any two partitions $\mathscr{P}_i, \mathscr{P}_j \in \mathfrak{P}$ the condition:
$f_h(\mathscr{P}_i) \leq f_h(\mathscr{P}_j)$ if and only if $\mathscr{P}_i \vartriangleleft \mathscr{P}_j$

Having summarized the necessary definitions we can now proceed by describing the basic HC algorithm.

5.3 Distances in Classical Hierarchical Clustering

We will now describe the basic HC algorithm and the distances that influence its outcome. Since it was first mentioned in this form by Ward in 1963 [95], many different kinds of HC schemes have been introduced. They are divided in agglomerative schemes in which clusters are iteratively united, and divisive schemes in which they are iteratively divided [92, 96, 94]. Agglomerative clustering is computationally more efficient and more flexible, hence, we will focus

5.3 Distances in Classical Hierarchical Clustering

on the classical agglomerative scheme proposed by Ward. Although this is a well known procedure we describe it in Algorithm 5.1 to illustrate how it should be used in the context of THz-TDS spectra and how the different parameters that can be chosen may be influenced.

Algorithm 5.1: Classical Hierarchical Clustering [95]

Data: Set of objects: $O = \{o_i\}_{i=1}^n$, with $n \in \mathbb{N}$,
Result: Hierarchy $\mathscr{H} := \{\mathscr{P}_i\}_{i=1}^h \subset \mathfrak{P}_O$ as defined in 5.4
Initial index set $M_1 := M$
Initial partition: $\mathscr{P}_1 := \{C_i\}_{i \in M_1}$, where $C_i = \{o_i\}$, for all $i \in M_1$
Calculate initial distance matrix $D_{\mathscr{P}_1}$:

1 **for** $i, j \in M_1$ **do**
  $\quad D_{\mathscr{P}_1}(i,j) := dist(o_i, o_j)$

Set Counter $h := 1$
while $|M_h| \neq 1$ **do**
  Find $p, q := \underset{p,q \in M_h, p \neq q}{\mathrm{argmin}}\; D_{\mathscr{P}_h}(p, q)$
2 Set $f_h(\mathscr{P}_h) := D_{\mathscr{P}_h}(p, q))$
  Update:
  Cluster $C_q^{new} := C_p \cup C_q$ Partition $\mathscr{P}_{h+1} := (\mathscr{P}_h \setminus \{C_p, C_q\}) \cup C_q^{new}$
  Index $M_{h+1} := M_h \setminus \{p\}$
  Distance matrix: **for** $i, j \in M_{h+1}$ **do**
    if $i, j \in M_h \setminus \{q\}$ **then**
      $\quad D_{\mathscr{P}_{h+1}}(i, j) := D_{\mathscr{P}_h}(i, j)$
    else
3   Calculate distance between the new cluster and all old clusters:
      $\quad D_{\mathscr{P}_{h+1}}(q, i) := dist(C_q^{new}, C_i)$
  Increment: $h := h + 1$;

$\mathscr{H} := \{\mathscr{P}_1, ..., \mathscr{P}_h\}$

Remark 5.6 Remembering the clustering index in Step 2 is not mandatory. It serves interpretation purposes. It is not necessary, however, for the clustering result.

5 Automatic Segmentation of Spectral Data

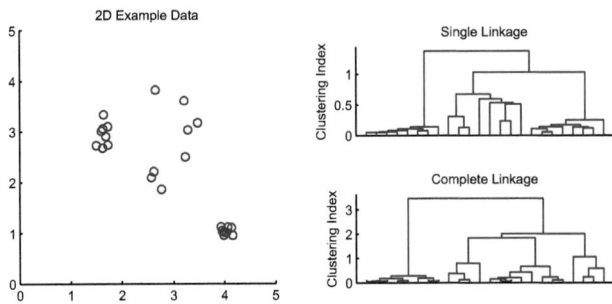

Figure 5.1: Visualization of HC.

The termination criterion of Algorithm 5.1 is given by the finiteness of the set. To illustrate how one can visualize a HC result, Fig. 5.1 shows a simple two-dimensional data set on the left hand side and two different hierarchies on the right hand side. The samples consist of 3 independently generated sets, i.e. in the best case three clusters should be detected. One of which is very dense, one has a a high variance and a third one a non-round shape. On the right hand side two examples of a typical visualization are displayed, so-called dendrograms. In [68] the author differentiates between two forms of dendrograms: Threshold dendrograms and proximity — in our case distance — dendrograms that display not only the branching but also the respective clustering index. As already mentioned in remark 5.6, remembering the clustering index is not necessary for the branching structure, and a threshold dendrogram is entirely drawn without this information. Fig. 5.1, however, shows distance dendrograms, as they are often easier to interpret. The clustering index on the y-axis can be used to find a cut-off level that provides good separation properties between the clusters.

A common criticism on HC algorithms is that the number of clusters has to be fixed at some point and that the problem of finding this parameter is only shifted in comparison to PC. However, changing the number of clusters in PC usually implies a complete recalculation, in HC it just means picking another one of the already calculated partitions. The algorithm itself does not directly depend on parameters.

5.3 Distances in Classical Hierarchical Clustering

The outcome of HC is not purely automatic and can be influenced. It is apparent that the decision to merge two clusters to a new one depends on their distance to each other. And if two clusters are merged this distance is recalculated. Consequently HC is influenced by two forms of distances:

(1) The distance between the samples calculated in Step 1 of Algorithm 5.1. It determines the first initial distance matrix, denoted by $D(o_i, o_j)$ for all o_i, o_j in the object space O

(2) The distance between a newly formed Cluster and all other clusters calculated in Step 3 of Algorithm 5.1, denoted by $D(C_p \cup C_q, C_i)$ for all i in the set $M \smallsetminus \{q, p\}$

We will now analyze some of the possibilities how to calculate these distances.

5.3.1 Distances between samples

Closeness between objects can be measured in various ways. Generally, in HC, some algorithms use distance measures and others use similarity measures. We will now define the requirements for both. The basis of calculating either is always the representation of the data set O as defined in 4.1, i.e. based on $X := a(O) \subset \Omega$.

Definition 5.7 For arbitrary feature sets Ω a function $d : \Omega \times \Omega \to \mathbb{R}^+$ is called a **distance** measure on Ω if:

(1) $d(x,x) = 0$ for all $x \in \Omega$ (positivity)

(2) $d(x,y) = d(y,x)$ for all $x,y \in \Omega$ (symmetry)

Definition 5.8 For arbitrary sets Ω a function $s : \Omega \times \Omega \to (-\infty, 1]$ is called a **similarity** measure on Ω when

(1) $s(x,x) = 1$, for all $x \in \Omega$ (positivity)

(2) $s(x,y) = s(y,x)$ (symmetry)

Remark 5.9 These definitions of similarity and distance are dual, i.e. one can easily transform similarities into distances by applying $f : \mathbb{R}^+ \to (-\infty, 1]$ with $t \mapsto 1 - t$, and vice versa. In this work mostly distances will be used. Because of the duality they can easily be transformed to be applicable on similarities.

5 Automatic Segmentation of Spectral Data

The above described distance measure definition does not concur with the definition of a metric. It only represents the two properties that are most frequently required by clustering algorithms. Nevertheless, in $\mathbb{R}^q$ the metrics induced by p-norms are frequently used. These distance measures additionally hold the triangle inequality $d(x,y) + d(y,z) \geq d(x,z)$ and strict positivity i.e. $d(x,y) = 0 \Leftrightarrow x = y$.

In this work we will employ two different distance measures in combination with the basic algorithm. The most commonly used measure in clustering is the Euclidean distance:

$$d : \mathbb{R}^q \times \mathbb{R}^q \to \mathbb{R}^+, \text{ with } d(x,y) = \sqrt{\sum_{i=0}^{q}(x_i - y_i)^2} \qquad (5.1)$$

The Euclidean distance is popular for various reasons, especially for its property of emphasizing great distances and suppressing smaller ones. Furthermore, with this metric analytically finding the centroid of a cluster can be done be just calculating the mean, which is not possible with the Manhattan distance ($p = 1$) for example.

Apart from the good characteristics of the Euclidean distance, there are applications where other distance measures perform better. One should therefore account for the specific application when the distance measure is chosen. In our case, the application are THz-TDS spectra as described in Chapter 2. The focus of interest in this application are the peaks, therefore, an appropriate distance measure would enhance the similarity in the peak position and depth while at the same time suppressing the differences in overall absorption. The baseline correction of Section 2.3.3, was done partly to suppress differences in the overall absorption. It is one method to produce a constant baseline for transmittance spectra. But such a baseline correction might not be possible because of reasons of computational efficiency and poor data quality. Additionally, after the baseline correction there can still be artifacts that lead to differences in the peak expression of different measurements of the same compound. To illustrate the problem with the varying overall absorption, in Fig. 5.2 uncorrected transmittance spectra are shown. With increasing frequency the baseline as well as the peaks expression becomes more variable. This is partly due to the logarithmic dependence of the transmittance ($\propto$ absorption coefficient as defined in Equation 2.1) from the magnitude spectra of the ratio of the sample and the reference. Therefore, although the spectra in the figure all belong to the same compound,

5.3 Distances in Classical Hierarchical Clustering

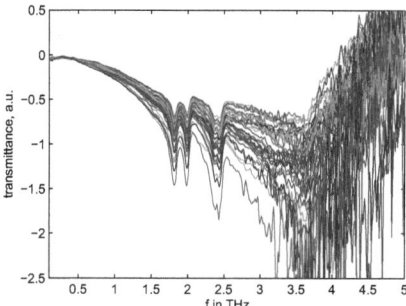

Figure 5.2: Variance of THz spectra.

the Euclidean distance accounts for the differences in higher frequencies in the same way as for the difference in the lower ones. A better distance measure would put a higher emphasis on the peak's positions and depth and suppress the effect the overall distance.

What we would need for this purpose is a distance measure that is invariant to vertical shifts and at the same time spectra that are normalized in a way that the information that should not be corrected is close to zero.

There are different ways to produce a shift invariant distance measure. However, in a different context, namely document clustering, the most popular similarity measure used has this property. It is called cosine similarity. The feature space Ω is considered a usual vector space and the indicator for the similarity of two vectors in the cosine of the angle between two such vectors [97], with $\Omega = \mathbb{R}^q$ this is:

$$s : \mathbb{R}^q \times \mathbb{R}^q \to [-1,1] \subset \mathbb{R}, \text{ with } s(x,y) = cos(x,y) = \frac{<x,y>}{\|x\|_2\|y\|_2} \quad (5.2)$$

As we need a distance measure, we will use the transformation as described in remark 5.9, i.e.

$$d : \mathbb{R}^q \times \mathbb{R}^q \to [0,2] \subset \mathbb{R}^+, \text{ with } d(x,y) = 1 - cos(x,y) = 1 - \frac{<x,y>}{\|x\|_2\|y\|_2} \quad (5.3)$$

Before applying such a distance measure the information that should not be altered has to be set to zero. If the spectra are baseline corrected, the baseline

5 Automatic Segmentation of Spectral Data

just has to be set to zero. For uncorrected transmittance spectra we propose to normalize the spectra such that the maximal transmittance is set to zero. As one can see in Fig. 5.2 the high transmittance regions are usually quite well concentrated already and the shift invariance therefore has a greater effect on the high frequency region. While the Euclidean distance would produce big distances between two spectra of Fig. 5.2 such a distance measure would have a tendency to suppress the unwanted distances close to the noise floor.

This distance measure will be applied and compared to the Euclidean distance measure. The cosine distance only satisfies the definition of a distance — 5.7. It neither satisfies the triangle inequality nor strict positivity [97].

5.3.2 Distances between Clusters

Distances between clusters are usually based on the distances between the objects. Typical approaches consist in using the minimal, maximal or average distance between the objects in the two clusters to define the distance between the classes. Many of the functions that can be used here were summarized in 1966 in the Lance-Williams formula [98]. In the above described HC a new cluster is build by uniting two old clusters. In the Lance-Williams formula the distance between a new class and all others depends directly on the distances the old clusters had to the other clusters. Thereby, all $D_{\mathscr{P}_i}$ can be traced back to $D_{\mathscr{P}_1}$. The initial distance matrix is equal to the distance matrix of the objects themselves. The latter was the topic of the previous section, i.e. $D_{\mathscr{P}_1}(i,j) := dist(o_i, o_j)$ for all $o_i, o_j \in O$.

The cluster distance can hence be characterized in the following way. Given a partition $\mathscr{P} \in \mathfrak{P}_O$ with $\mathscr{P} := \{C_j\}_{j \in M}$, the distance between a newly formed cluster $C_p \cup C_q$, $p, q \in M$ and all other clusters is defined by:

$$D(C_p \cup C_q, C_i) := \alpha_p D(C_p, C_i) + \alpha_q D(C_q, C_i) + \beta D(C_p, C_q) + \gamma |D(C_p, C_i) - D(C_q, C_i)| \tag{5.4}$$

The coefficients α_p, α_q, β, and γ determine the type of linkage function. Although there are many more [94], in this work we will use the the following sets of coefficients:

(1) The parameter set:

$$\{\alpha_p, \alpha_q, \beta, \gamma\} := \{\frac{1}{2}, \frac{1}{2}, 0, -\frac{1}{2}\}$$

5.3 Distances in Classical Hierarchical Clustering

Single linkage (SLink): Minimum of previous distances, i.e.

$$D(C_p \cup C_q, C_i) = min\{D(C_p, C_i), D(C_q, C_i)\}$$

(2) The parameter set:

$$\{\alpha_p, \alpha_q, \beta, \gamma\} := \{\frac{1}{2}, \frac{1}{2}, 0, \frac{1}{2}\}$$

Complete linkage (CLink): Maximum of previous distances, i.e.

$$D(C_p \cup C_q, C_i) = max\{D(C_p, C_i), D(C_q, C_i)\}$$

(3) The parameter set:

$$\{\alpha_p, \alpha_q, \beta, \gamma\} := \{\frac{1}{2}, \frac{1}{2}, \frac{D(C_p, C_i), D(C_q, C_i)}{2}\}$$

Average linkage: Average of previous distances, i.e.

$$D(C_p \cup C_q, C_i) = \frac{D(C_p, C_i) + D(C_q, C_i)}{2}$$

With these parameter sets one can discover different kinds of clusters. We will illustrate this on the two-dimensional application example from Fig. 5.1. In Fig.

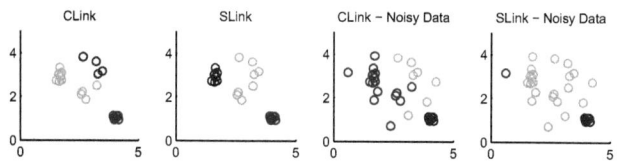

Figure 5.3: Different linkage functions applied on images without (left) and with (right) noise.

5.3 we see the example data set grouped with different linkage functions once

5 Automatic Segmentation of Spectral Data

without and once with noise. On the left hand side the first three clusters of this data set with the Slink and the Clink method are shown. The Slink detects the blue clusters — the long shape — better, as it does not assume globular shapes. On the right hand side we see the same data set with some equally distributed noise added. Here the Clink detects almost the same clusters as before while the Slink clustering is influenced by the noise.

We have now given a first introduction into the basic HC algorithm and the possibilities for the user to influence the classification. We introduced a distance measure that is usually applied in document retrieval to time-series clustering and reasoned its adequateness for the special application of THz-TDS. There are still some drawbacks of the classic algorithm which can not be overcome by changing distance functions. To overcome or diminish these drawbacks, we now introduce an algorithm that was more recently developed and analyze its properties as well as propose some alterations.

5.4 Discussion of Karypis Chameleon Algorithm

The overall computational complexity of Algorithm 5.1 is $O(n^2)$ in time as well as in space [96]. With high data volume the computation might fail because the system is out of memory. Even a moderately sized hyperspectral image with 256× 256 pixels for example contains about 65000 samples (pixels, i.e. spectra). And a distance matrix of double values — which has to be updated continuously in HC, i.e. has to be held in main memory — would need 4 Gigabyte. The quantity of data of such images exceeds the capacity of many computers. Therefore, especially in the late 1990s a number of different algorithms have been developed that bypass this problem by pre-clustering or sub-sampling. The most popular [92, 99] of these algorithms are Birch [100], Cure [101] and Chameleon [102]. In Birch the entire data is organized in a tree structure in one scan and the final clustering is executed only on the representatives (means) of the leaf node groups. The Cure algorithm uses a randomly sampled and then partitioned sub-sample of the original data. The clustering basis is not just one representative but a well scattered number of representative points. In the Chameleon clustering a k-nearest neighbor graph is build with the whole data and then cut into clusters with a minimum edge-cut bisector.

All these algorithms have in common that the final hierarchical clustering is not carried out on the full data set but on a subset or transform. Therefore

5.4 Discussion of Karypis Chameleon Algorithm

the computational time mainly depends on the size of this subset and on the efficiency of the down-sampling method. We decided to use the full Chameleon algorithm as well as only the pre-clustering of this algorithm. To explain why we chose this algorithm we will first describe it in Algorithm 5.2. As the Chameleon is based on a nearest neighbor network, this is a necessary definition we have to give beforehand:

Definition 5.10 We define the **k-nearest neighbor network** as a network (G, w) with an undirected graph $G := (X, E)$ and the weights $w : E \to \mathbb{R}^+$ that satisfies:

(1) Each vertex $x \in X$ appears in at least k pairs $\{x, \cdot\}$ in E.

(2) All $e \in E$ are unordered pairs $e = \{x, y\}$.

(3) For all $e = \{x, y\} \in E$ the weight is defined as a similarity (Definition 5.8), i.e. $w(\{x, y\}) := s(x, y) = s(y, x)$. And if $\{x, y\} \in E$ it follows that $s(x, y) \geq s(x, z)$ for all $z \in X$ with $\{x, z\} \notin E$.

Remark 5.11 In the Chameleon algorithm everything is denoted with similarities instead of distances, we will therefore also talk about closeness from the similarities point of view in this section. The definitions and nomenclature follows standard graph-theory nomenclature. For a deeper insight into this matter we refer the reader to the vast literature on discrete mathematics, e. g. [103].

Chameleon can be divided in two phases. In the first one the initial partition is build, as explained in Step 1 and Step 2. In the second one, Step 3, the final clustering is performed. The main difference between the named efficient hierarchical clustering algorithms and Algorithm 5.1 is the method used for preclustering, but in contrast to Birch and Cure the Chameleon pre-clustering is very easily implemented, mainly because it is based on well known algorithms.

First, in Step 1 the nearest neighbor network as explained in Definition 5.10 has to be calculated. As nearest neighbor graphs are useful in many applications, there are efficiently implemented open source libraries one can use for this purpose. We chose kdtree2 by Matthew B. Kennel [104]. This implementation finds the k-nearest neighbor graph in $O(log(n))$ time, which is an improvement to the brute-force implementation which takes $O(n)$. The storage depends on k because for each sample the k nearest neighbors and the distance to them have to be stored.

5 Automatic Segmentation of Spectral Data

Algorithm 5.2: The Basic Procedure of the Chameleon Algorithm [102]

Data: Set of samples $X := a(O)$ with $X \subset R^q$ and $O = \{o_i\}_{i=1}^n$, with $\{1,...,n\} =: N \subset \mathbb{N}$, number of nearest neighbors k and number of initial clusters m.

Result: Hierarchical partition of data $\mathscr{H} := \{\mathscr{P}_i\}_{i=1}^h \subset \mathfrak{P}_O$ as defined in 5.4

1 Build a k-nearest neighbor network of the sample set X
2 Build a partition $\mathscr{P}^{\text{Pre}} = \{C_i\}_{i \in M}$ of the k-nearest neighbor graph parts by minimizing:
$$ES(\mathscr{P}^{\text{Pre}}) = \sum_{e \in E \setminus \bigcup_{i \in M} E|_{C_i}} w(e) \qquad (5.5)$$
with $E|_{C_i} := \{e = \{x,y\} \in E | x \in C_i \wedge y \in C_i\}$
3 Apply Algorithm 5.1 with $\mathscr{P}_1 := \mathscr{P}^{\text{Pre}} = \{C_i\}_{i \in M}$ and two special distance measures that are based on the nearest neighbor graph

---

Next, in Step 2 the initial partition is build from the network. Also for this purpose we chose an already available library, namely the 2007-Version of Karypis METIS library [105]. The problem of partitioning a graph is NP-complete, nevertheless with the algorithms implemented in this library one can find good approximations of minimum edge-cut partitions in reasonable computational time [106]. It has been successfully compared to other m-cut libraries such as Jostl and PaGrid [107].

The availability of efficient algorithm for the implementation of Phase I is, one of the reasons why we chose the Chameleon algorithm. We can either use the full procedure or combine Phase I with Algorithm 5.1. There is a version of Chameleon implemented in the software package CLUTO [108], however, as we want to have the possibility to combine Phase I with classical hierarchical clustering and its different parameters, we chose to re-implement the algorithm.

Remark 5.12 Two parameters have to be set in Phase I, the number of nearest neighbors k and the number of initial clusters m. The algorithm is not very sensitive to this choice. Generally, the higher the number of nearest neighbors the worse the computational efficiency and the lower the number of initial clusters the more likely it is that important information is lost. Therefore, we generally choose $k := \sim 0.05 \times n$ and $m := \sim 0.1 \times n$.

5.4 Discussion of Karypis Chameleon Algorithm

The other reason for choosing this algorithm is that Phase II, i.e. Step 3, is based on a distance function that includes the whole nearest neighbor information into the further clustering. It is said to give results that are less sensitive to noise than the classical hierarchical clustering is. We will now explain the similarity measures that are used in Step 3 in the way they are explained in [102]. We will then proceed to illustrate some problems that may occur when using these similarities and we will propose alterations of this method that can be used to overcome these problems.

5.4.1 The Distance Measure

Two different functions are combined for a similarity measure that covers clusters of different shape and density. They are called the relative closeness RC and the relative inter-connectivity RI. We will analyze them mathematically and highlight some of their properties in that respect.

Definition 5.13 Given a nearest neighbor network (G, w) of a graph $G := (X, E)$ and a minimal m-cut partition of X in $\mathscr{P}^{\text{Pre}} := \{C_i\}_{i \in M}$ to m equal parts we define the **relative inter-connectivity** (RI) between two clusters by

$$RI(C_i, C_j) = \begin{cases} \frac{\|EB_{C_i,C_j}\|}{\frac{\|EW_{C_i}\| + \|EW_{C_j}\|}{2}} & \text{for } EB_{C_i,C_j} \neq \emptyset \\ 0 & \text{else} \end{cases} \quad (5.6)$$

where the EB and EW are subsets of E; the former are the edges between (i.e. EB) clusters and the latter the edges a minimum edge-cut bisection within (i.e. EW) a cluster would contain.

(1)

$$EB_{C_i,C_j} := \begin{cases} EW_{C_i} & \text{for } C_i = C_j \\ \{e \in (E|_{C_i \cup C_j} \setminus (E|_{C_i} \cup E|_{C_j}))\} & \text{else} \end{cases} \quad (5.7)$$

(2) EW_C denotes the set of edges a minimum edge-cut bisection of the subgraph $(C, E|_C)$ into C_1 and C_2 would have (minimizing an equation analogous to 5.5). That means $EW_C := \{e \in (E|_C \setminus (E|_{C_1} \cup E|_{C_2}))\}$

(3) For $E' \subseteq E$ the term $\|E'\|$ is defined as the sum of the weights of the edges, i.e. $\|E'\| := \sum_{e \in E'} w(e)$

5 Automatic Segmentation of Spectral Data

Definition 5.14 The **relative closeness** (RC) is defined as a measure of the average distance between clusters.

$$RC(C_i, C_j) = \begin{cases} \frac{|C_i|}{|C_i|+|C_j|}\overline{EW_{C_i}} + \frac{|C_j|}{|C_i|+|C_j|}\overline{EW_{C_j}} & \text{for } EB_{C_i,C_j} \neq \emptyset \\ 0 & \text{else} \end{cases} \quad (5.8)$$

for $E' \subseteq E$ the term $\overline{E}$ is defined by $\overline{E'} := \frac{\|E'\|}{|E'|}$, i.e. denotes the average of the weights of edges in E'.

The similarity is then calculated by $S(i,j) := RI(C_i, C_j) RC(C_i, C_j)^\alpha$, where α controls the importance of the closeness in relation to the inter-connectivity.

Remark 5.15 The **initialization** of the algorithm described in Section 5.3 is done by $\mathscr{P}_i := \mathscr{P}^{\text{Pre}}$ and $M_1 := M$. The initial similarity matrix is defined by $S_{\mathscr{P}_1} := S$. For updating purposes EB (a symmetric $m \times m$ matrix) and EW (a m vector) are also stored.

Remark 5.16 With analogous nomenclature as in Section 5.3 the clustering is carried out. One difference is, of course, that not the clusters with the smallest distance but the ones with the biggest similarity are united. In step 7 EB and EW have to be updated to get the correct similarity matrix S. One can use the old edge-cut information for $EB_{C_p \cup C_q, C_i} = (EB_{C_p, C_i} \cup EB_{C_q, C_i}) \setminus EB_{C_p, C_q}$. The set $EW_{C_p \cup C_q}$ however can not be determined from previous information. One has to recalculate the minimum edge-cut bisector as one of the conditions for such a bisector named in the paper is, that the sections should be approximately equally sized. This is not necessarily the case for C_p and C_q, hence $EW_{C_p \cup C_q}$ has to be recalculated in each step.

The similarity function used within the Chameleon algorithm yields good results in general but using it during the clustering process can produce several problems. In the next sections we will analyze two of these flaws and propose solutions for them.

5.4.2 Alteration: Changing the Similarity Measure

The Chameleon similarity term $S(i,j)$ is not a similarity as defined in 5.8. The problem is the closeness term RC which is not necessarily smaller than 1. We

5.4 Discussion of Karypis Chameleon Algorithm

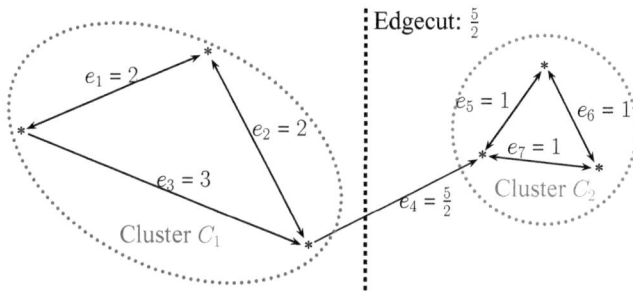

Figure 5.4: Counter example for $RC \leq 1$.

give a counter example in Fig. 5.4. The figure shows a 2-nearest neighbor graph with 6 edges. To illustrate the inherent non-symmetry of a nearest neighbor graph, it is depicted as a directed graph. For the further calculations however, the graph is assumed not to be directed, i.e. every vertex counts one time. When cutting the graph at e_4, one gets a minimum edge-cut for a 2-way partition. Therefore $EB_{C_1,C_2} = \{e_4\}$, EW_{C_1} must be $\{e_1, e_2\}$ and for EW_{C_2} there are several possibilities, e. g. $\{e_5, e_6\}$. Hence, we have: $RI(C_1, C_2) = \frac{\frac{5}{2}}{\frac{4+2}{2}} = \frac{5}{6}$ and $RC(C_1, C_2) = \frac{\frac{5}{2}}{\frac{3\cdot 4}{6\cdot 2} + \frac{3\cdot 2}{6\cdot 2}} = \frac{5}{3}$. Not only is $RC(C_1, C_2) \geq 1$ in this example but additionally with $\alpha = 1$ the similarity $S(1, 2) = \frac{25}{18} \geq 1$. Thus, neither RC nor S are similarities as defined in Definition 5.8.

Using a closeness indicator not satisfying the similarity measure conditions has one major drawback: Objects can be more equal to another than one object is to itself. This is usually not wanted and would be expected to produce biased results. Therefore we propose a slightly improved calculation of the similarity S. This works because RI is a similarity to begin with and one can proof that RC and therefore also S are bounded by the size of the nearest neighbor graph.

First we show that RI is a similarity. $RI : \mathscr{P}_1 \times \mathscr{P}_1 \to [0, 1]$, because on the one hand one can easily see that $RI(C_i, C_j) \geq 0$ for all i, j. On the other hand if there were i, j such that $RI(C_i, C_j)$ was bigger than 1 then EB_{C_i, C_j} would be smaller than $EW_{C_i} + EW_{C_j} + EB_{C_i, C_j}$. This is a contradiction to C_i, C_j belonging

5 Automatic Segmentation of Spectral Data

to a minimum edge-cut partition (because the cut $C_{i_1} \cup C_{j_1}$ and $C_{i_2} \cup C_{j_2}$ would have had a smaller weight sum). Positivity and symmetry are given by definition.

The term in question is RC and thereby also S. We now show that RC is bounded, because w.l.o.g. let $|C_i| \leq |C_j|$, and $|EW_{C_i}| \leq |EW_{C_j}|$, then

$$
\begin{aligned}
RC(C_i, C_j) &\leq \frac{\overline{EB_{C_i,C_j}}}{(EW_{C_1}+EW_{C_2})\frac{|C_i|}{|C_i|+|C_j|}} \\
&\leq \frac{\|EB_{C_i,C_j}\|}{\|EW_{C_i}\|+\|EW_{C_j}\|} \left(\frac{|EW_{C_i}|}{|EB_{C_i,C_j}|}\right) \frac{|C_i|+|C_j|}{|C_i|} \\
&\leq \frac{1}{2}\left(\frac{|EW_{C_i}|}{|EB_{C_i,C_j}|}\right) \frac{|C_i|+|C_j|}{|C_i|} \\
&\leq \frac{1}{2}\left(\frac{k|C_i|}{1}\right) \frac{|C_i|+|C_j|}{|C_i|} \\
&= \frac{k(|C_i|+|C_j|)}{2}
\end{aligned}
$$

This hold for all i, j, where k is the number of nearest neighbors. We now substitute RC from Equation 5.8 by the following:

$$
RC(C_i, C_j) = \begin{cases} 1 & \text{for } i = j \\ \frac{2}{k}\left(\frac{\overline{EB_{C_i,C_j}}}{|C_i|\overline{EW_{C_i}}+|C_j|\overline{EW_{C_j}}}\right) & \text{for } EB_{C_i,C_j} \neq \emptyset \\ 0 & \text{else} \end{cases} \qquad (5.9)
$$

With these alterations RC, RI and S are similarity measures. Furthermore, instead of recalculating RI and RC in every step, we calculate the similarity matrix $\hat{S} = RI RC^\alpha$.

5.5 Simplified Chameleon

The second problem that can arise when the Chameleon algorithm is executed is that the updating function can produce so-called inversions or crossovers, i.e. the index (as defined in 5.5) is not monotone. A crossover would then produce a higher similarity / lower distance by updating the matrix than appeared in it before. There are other hierarchical clustering algorithms that do produce inversions such as the centroid linkage and the median linkage method [94, 68]. This effect does not imply that no hierarchy can be build. The clustering itself can still work well and produce satisfying results. However, inversions can impede the interpretation of the hierarchy because no distance dendrogram can be drawn. Hence, the distance information can not be used to find an appropriate cutting level.

5.5 Simplified Chameleon

For the above mentioned reasons we propose to alternatively use the pre-clustering of the Chameleon algorithm and then combine the result of this pre-clustering with the classical hierarchical clustering method. We describe this in Algorithm 5.3

Algorithm 5.3: Proposed Simplified Chameleon Algorithm

Data: Set of samples $X := a(O)$ with $X \subset R^q$ and $O = \{o_i\}_{i=1}^{n}$, with $n \in \mathbb{N}$, number of nearest neighbors k and number of initial clusters m.

Result: Hierarchical partition of data $\mathcal{H} := \{\mathcal{P}_i\}_{i=1}^{h} \subset \mathfrak{P}_O$ as defined in 5.4

Build a k-nearest neighbor network // **Analogous to Algorithm 5.2**
Build initial partition $\mathcal{P}^{\text{Pre}} = \{C_i\}_{i \in M}$ with k-nearest neighbor graph
// **Analogous to Algorithm 5.2**

1 Calculate the distance matrix between the initial clusters C_i
2 Apply Algorithm 5.1 with $\mathcal{P}_1 := \mathcal{P}^{\text{Pre}} = \{C_i\}_{i \in M}$ and typical linkage functions

---

The first steps works as explained in Algorithm 5.2. In Step 1 we propose two alternative approaches to gain an initial distance matrix:

(1) Calculate a representative $\overline{x_i}$ for each $C_i \in \mathcal{P}^{\text{Pre}}$, by using the mean of the members of C_i, i.e. $\overline{x_i} := \sum_{x \in C_i} \frac{x}{|C_i|}$. Calculate the initial distance matrix by calculating the distances between the means of the clusters $\overline{x_i}$ and using arbitrary distance functions, e.g. the ones explained in Section 5.3.2.

(2) Use $\hat{S}$ as defined in the previous section as the basis for classical hierarchical clustering. i.e. instead of recalculating the distances on the basis of the nearest neighbor graph in every step we use the linkage functions as described in Section 5.3.2. The advantage of this methods is that the method still contains information about the underlying densities and closeness of the clusters, although less, but the clustering algorithm does not fail and can provide information about distances within the hierarchy if necessary.

5 Automatic Segmentation of Spectral Data

The main difference between Algorithm 5.3 and the classical Chameleon algorithm is that the distance between the samples are not recalculated in Step 2 of each iteration. There rather is an initial distance matrix used and then updated with the classical linkage functions that were described in 5.3.2.

The disadvantage is that the two aspects of closeness and density are joined together again. The advantage however, is its flexibility in the use of updating functions and the stability with respect to inversions.

5.6 Spatial Domain

Up until now the automatic organization of the spectra is done on the basis of spectral characteristics only. That means that we cluster the samples assuming that they are acquired independently from each other.

This entails adaptability to rather large incrementally acquired data sets. Often the data is not acquired independently, though. And by the previously discussed organization, no dependency information is included. In hyperspectral imaging the samples often belong to a spatially coherent object and therefore are far from independent. Nevertheless, the use of spatial features in segmentation of hyperspectral images is often neglected [109] although it is likely that this information can improve the segmentation quality.

We will now discuss two different approaches that are applied when spatial information is included. The first one is classical two-dimensional image processing on the feature reduced representation of the hyperspectral channels. The other one is the direct inclusion of neighborhood information into the clustering algorithm.

5.6.1 Classical Image Processing

In many hyperspectral imaging applications large regions of the image can quickly be discarded from further analysis. Especially in applications where hidden chemicals need to be detected, only few areas will potentially contain such chemicals. Particularly areas that do not contain any absorbing information as well as areas that contain completely absorbing material can usually be identified quite easily. The focus should lie on the material in between, which is not completely absorbing but contains only some absorption peaks. The first step therefore, should be to identify these regions of interest. If this is done, the respective areas can be

5.6 Spatial Domain

further denoised by smoothing functions.

A variety of two-dimensional smoothing algorithms can be applied to improve the channel information. A popular one is the anisotropic diffusion filter. It offers very good smoothing properties while at the same time preserving the edges. This filter has been applied on hyperspectral images by different researchers. [110] and [109] apply it slice-wise, i.e. the two-dimensional version of the filter. For reasons of computational feasibility, they do not use the full data set but rather sub-samples that are based on a kind of PCA. In contrast to that, [111] uses the full hyperspectral images to calculate a $2\frac{1}{2}$D diffusion filter that smooths spatially on the slices and at the same time spectrally on the time-series. In Chapter 2 we considered spectral smoothing techniques that can be used for peak preserving smoothing in the spectral domain, e. g. Savitzky-Golay filtering, wavelet shrinkage, and more specifically echo removal. Therefore, we do not apply any more spectral smoothing here but decided to include spatial characteristics by slice-wise smoothing. Referring to the introduction of this chapter we propose not to use the full spectra but the PFS from Excursus 2. They are the basis for 2D image processing we propose in Algorithm 1. In Step 1 of the algorithm we perform a first segmentation on the basis of those two feature images that contain the best overall information about the whole image content, namely PFS 1 and 6. The segmentation is performed with a k-means clustering [90, 70] that was briefly explained in Section 5.1. In Step 2 different possibilities for a smoothing method can be used. Valid results can be achieved with anisotropic diffusion [112], or bilateral filtering [113].

There is an alternative to level dependent smoothing. It consists of calculating one smoothing function that is then applied to all levels. One possibility is using an edge preserving bilateral filter that depends on the spectral distance of the pixels [113]. The smoothed image I_{Sm} from a bilateral filter is calculated in each pixel p by:

$$I_{\text{Sm}}^{k}(p) = \frac{1}{W} \sum_{q \in S} G_{\sigma_s}(\|p - q\|) G_{\sigma_t}(|I^k(p) - I^k(q)|) I^k(q) \qquad (5.10)$$

where W is a normalization term, G_{σ_s} and G_{σ_t} are two Gaussian kernels that influence how much spatial information s is included (represented by the pixel positions p and q, S is the neighborhood that is determined by σ_s) and how much range information t. The range information can be calculated level-wise, i.e. considering only the values in each level image I^k, i.e. $|I^k(p) - I^k(q)|$ is the distance between the level values in p and q. But it can also be calculated

5 Automatic Segmentation of Spectral Data

Algorithm 5.4: Spatial Similarity Inclusion with Slice-wise Image Processing

Data: Hyperspectral $p \times q$ pixel Image with pixel set $X := \{x_i\}_{i=1}^{pq}$ where x_i are THz transmittance spectra
Result: Hierarchical clustering of image
for $x_i \in X$ **do**
$\quad$ Calculate PFS representation X_{PFS}

1 Retrieve segmentation into three parts from PFS feature 1 and PFS feature 6
Categorize segments into: Highly absorbing X_H, non-absorbing X_{No}, and medium absorbing X_M
for *PFS coefficients* $j = 1$ **to** E **do**

2 $\quad$ Apply an edge preserving smoothing filter on $X^*_{\text{PFS}_i}$
$\quad$ Mask all PFS level images with X_M:

$$X^*_{\text{PFS}_j} := \{x^*_{\text{PFS}_{ji}}\}_{i=1}^{pq}, \text{ where } x^*_{\text{PFS}_{ji}} := \begin{cases} x_{\text{PFS}_{ji}} & \text{if } i \in M \\ 0 & \text{else} \end{cases}$$

Set $X_F := \{x_{F_i}\}_{i=1}^{pq}$ where $x_{F_i} := \begin{pmatrix} x^*_{\text{PFS}_{1i}} \\ \ldots \\ x^*_{\text{PFS}_{Ei}} \end{pmatrix}$

for $x_{F_i} \in X_F$ **do**
$\quad$ Apply clustering, e.g. Algorithm 5.1 or 5.2 to retrieve hierarchical decomposition of index set I and thereby X

5.6 Spatial Domain

from the feature vector distance. The second Gaussian kernel G_{σ_t} is then applied on some distance function $\|\cdot\|$ between the feature vectors $F(p)$ and $F(q)$ with $F(p) = (I^1(p), I^2(p), ..., I^f(p))$. Equation 5.10 thereon is substituted by:

$$I^k_{\text{Sm}}(p) = \frac{1}{W} \sum_{q \in S} G_{\sigma_s}(\|p-q\|) G_{\sigma_t}(\|F(p) - F(q)\|) I^k(q)$$

Algorithm 1 and application of the bilateral filtering on hyperspectral images outline two different ways to apply classical image processing on hyperspectral images. In the next section we will propose a method that includes spatial information directly into the clustering algorithm.

5.6.2 Distance Matrix Inclusion

The previously explained approaches have in common that they are applied on the feature data before the segmentation is performed. The feature vectors that are then used for the clustering might be smoothed in the spatial domain. Still the clustering itself and the cluster distances are, again, calculated as if the samples where independently acquired. For this reason, we propose to include a spatial similarity into the clustering algorithm itself. That means combining the spatial closeness with the range closeness in all distance calculations. The problem of applying bilateral filtering is, that as soon as clusters are formed, defining a neighborhood S is not trivial. When we work with both forms of the Chameleon algorithm explained in Chapter 5. The clustering is not carried out on the pixels but on groups of pre-clustered pixels. Bilateral filtering is not applicable in that case. We therefore have to find a similarity matrix that includes the neighborhood information into the distance calculations between these pre-clusters. For this purpose we propose to use a Gaussian approach that will be explained in the next paragraph.

For each two pre-clusters C_i^{Pre} and C_j^{Pre} from the initial partition $\mathscr{P}^{\text{Pre}}$ we assume that a distance $D(C_i^{\text{Pre}}, C_j^{\text{Pre}})$ has been calculated. We now want to include spatial information into this distance calculation. For this purpose we assume that the two clusters are independently acquired normally distributed data sets. For reasons of simplicity we then analyze their characteristics separately in the two spatial domains x and y. Algorithm 5.5 explains the inclusion of this kind of spatial distance between clusters.

A next step could be to include this closeness also into the updating mechanism of the hierarchical clustering itself. However, this is not sensible in the case of

5 Automatic Segmentation of Spectral Data

Algorithm 5.5: Proposed Inclusion of Spatial Similarity into Cluster Distance
---
Data: Distance matrix D^{Pre} defined on $\mathscr{P}^{\text{Pre}}$, parameter λ
Result: Updated distance matrix $D^{\text{Pre}*}$
Calculate a spatial distance $D^{\text{Sp}} : \text{Pre}^{\text{Pre}} \to \mathbb{R}^+$ in the following way:
for *each two $C_i, C_j \in \mathscr{P}^{\text{Pre}}$* **do**
  Calculate spatial overlap $O^x(i,j)$ and $O^y(i,j)$ by:
  for *Spatial dimension x and y* **do**
    Calculate the cluster mean spatial positions μ_i, μ_j and variance σ_i, σ_j
    Calculate overlap $O(i,j)$ of two Gaussian curves determined by parameters (μ_i, σ_i) and (μ_j, σ_j)
  Define the spatial distance by:

$$D^{\text{Sp}}(C_i, C_j) = \begin{cases} \frac{1}{O^x(i,j) + O^y(i,j)} & \text{if } O^x(i,j) \text{ and } O^y(i,j) > \lambda \in [0.25, 1] \\ 0 & \text{if } i = j \\ 2 & \text{else} \end{cases} \quad (5.11)$$

  Define updated distance matrix by
  $D^{\text{Pre}*}(C_i, C_j) := D^{\text{Sp}}(C_i, C_j) * D^{\text{Pre}}(C_i, C_j)$
---

hierarchical clustering as this would lead to a constant increase of the distance between the clusters. In classical hierarchical clustering the distance function is not recomputed from the original data in each iteration. Each step only uses the previous distance information that has been provided by the sub-clusters. If Equation 5.11 would be used, the updated distance matrix would have nothing to do with the original data and be very sensitive to any kind of noise. Therefore we apply this alteration of the distance matrix only to the initial one and further proceed in the traditional manner.

5.7 Evaluation of Hierarchical Clustering

We have now introduced and analyzed different methods of segmentation or clustering of the given data. These methods require evaluation. Cluster evaluation even today is by no means exhausted. To judge the quality of an algorithm and its parameters, different aspects have to be considered. Unsupervised classification is generally applied to organize either large or high-dimensional unlabeled data sets. Hence, two major quality aspects are scalability on the one hand and the ability to deal with the curse of dimensionality on the other hand [62]. Thus a critical part of evaluating the clustering is judging if the algorithm works, i.e. if it provides any formally sensible result and how fast it is. This topic has been addressed in this work in the previous section where the Chameleon algorithm was introduced with its features. However, the quality of the result has to be assessed in some way too. The premise of applying clustering algorithms is that the type of classes that can appear is not known beforehand. Hence, judging the accuracy of a result is intricate. Otherwise supervised or semi-supervised learning would be used.

5.7.1 State of the Art Methods

There is a vast amount of research on the topic of cluster quality evaluation [114, 68, 94]. In this thesis we will coarsely outline the different possibilities to asses this issue and focus on one that is often used to evaluate specifically HC results, the so-called F-score. We will present its characteristics and outline some flaws and then propose an alternative approach. We will furthermore propose an efficient implementation for this new evaluation method.

Although cluster evaluation of partitional and hierarchical algorithms often differs, many evaluation methods for HC are derived from methods for PC. Therefore, we will initially give some examples for general methods. A more detailed description of the hierarchical methods follows later on.

Cluster evaluation may be classified into external and internal criteria. External criteria serve the purpose of comparing different partitions with each other. Different partitions can either arise from diverse clustering algorithms applied on the same data set (in [114] this would be called a relative criterion) or they can arise from a given ground truth that should be matched as adequate as possible by a specific clustering algorithm. Typically, internal criteria evaluate the result with a particular focus on the density within the retrieved clusters on the one hand and the clear separation between two clusters on the other hand. Examples for these are separation indices such as the Dunn index [115]. These criteria are particularly useful when evaluating the choice of parameters — as for example the number of clusters — in PC algorithms [114]. They are seldom used for HC evaluation but more frequently to find a good clustering level. In HC internal criteria are applied to determine the appropriateness of a dendrogram for a given distance matrix. Usually a cophenetic matrix is calculated as a representative for the gained hierarchical structure and then is correlated with the original data [68].

Regarding feature selection and different distance measures — as for example described in Section 5.3 — we would like to compare the outcome of clusterings on the basis of different distance matrices. Therefore, the above mentioned internal criteria for hierarchical clustering can not be applied. We will hence focus on external criteria. The majority of these external criteria are conceptually made for PC. Some of them compare the clustering result to a random clustering of the data. Huberts Γ-statistic does that for example by estimating how far the labeling is from a random labeling. Monte Carlo simulations are used to tackle the computational complexity of the problem. Another approach is to compare clustering results from different algorithms and define a similarity or distance between them. Examples for these algorithms are the Mirkin distance, the van Dongen distance, the Rand index, and the FScore ([116], [114], [68], [117]). All of the latter are based on "counting sets", or on a type of confusion table where two partitions $\mathscr{P} = \{C_i\}_{i \in M}$ and $Pa' = \{C'_i\}_{i \in M'}$ of the same sample set X are compared: The entries are defined as follows:

- TP: $\{\{x,y\} \subseteq X | \exists (i,j) \in M \times M^g : x, y \in C_i \cap C_j^g\}$, i.e. all sample pairs that are in the same cluster in partition $\mathscr{P}$ and partition $\mathscr{P}^g$.

5.7 Evaluation of Hierarchical Clustering

	positive	negative
positive	true positives (TP)	true negatives (TN)
negative	false positives (FP)	false negatives (FN)

Table 5.1: confusion table

- TN: $\{\{x,y\} \subseteq X | \nexists (i,j) \in M \times M^g : x,y \in C_i \cup C_j^g\}$, i.e. all sample pairs that are in different clusters in partition $\mathscr{P}$ and partition $\mathscr{P}^g$.

- FP: $\{\{x,y\} \subseteq X | \exists i \in M : x,y \in C_i \wedge \nexists j \in M^g : x,y \in C_j^g\}$, i.e. all sample pairs that are in the same cluster in the partition $\mathscr{P}$ but not in the partition $\mathscr{P}^g$.

- FN: $\{\{x,y\} \subseteq X | \exists j \in M^g : x,y \in C_j^g \wedge \nexists i \in M : x,y \in C_i\}$, i.e. all sample pairs that are in the same cluster in the partition $\mathscr{P}^g$ but not in the partition $\mathscr{P}$.

The cardinality of these sets is then used to calculate indices representing similarities or differences between partitions of the same data set. In particular, if a data set with a ground truth partition is known, it can be used these sets to numerically evaluate the quality of the calculated partition.

5.7.2 The F-Score

We will now give a description of one of these evaluating functions — the so called F-score. The F-score was originally introduced in the area of information retrieval by Jardine and Rijsbergen in 1971 [118]. It has been adapted to be generally applicable on HC and since has been one of the most popular method to compare HC results with a ground truth partition [119, 120].

Two different functions on the partition space contribute to the F-score: The **precision** and the **recall**. The precision is defined by:

$$\text{Pre}(\mathscr{P}, \mathscr{P}^g) := \frac{|TP|}{|TP \cup FP|}$$

Assuming $\mathscr{P}^g$ to be the ground truth, the precision represents the proportion of objects retrieved correctly in relation to all the retrieved objects. The recall is defined by:

$$\text{Rec}(\mathscr{P}, \mathscr{P}^g) := \frac{|TP|}{|TP \cup FN|},$$

5 Automatic Segmentation of Spectral Data

and it represents the relation between correctly retrieved objects to the correct number of objects that should be in this class. The F-score is a combination of the two functions whose importance is controlled by the parameter β. Setting $\beta = 1$ gives equal importance to both.

$$\mathrm{FSc}(\mathscr{P}, \mathscr{P}^g) := \frac{(\beta^2 + 1) \times \mathrm{Pre} \times \mathrm{Rec}}{\beta^2 \times \mathrm{Pre} + \mathrm{Rec}}$$

We have $\mathscr{P} = \mathscr{P}^g \Rightarrow \mathrm{Rec}(\mathscr{P}, \mathscr{P}^g) = \mathrm{Pre}(\mathscr{P}, \mathscr{P}^g) = 1$, furthermore $\mathrm{Pre}(\mathscr{P}, \mathscr{P}_{\mathrm{GT}}) > 0$ and $\mathrm{Rec}(\mathscr{P}, \mathscr{P}_{\mathrm{GT}}) > 0$. Hence, FSc is a similarity measure on the partitions of X as defined in Definition 5.8. Thereby, on can define a function for comparing partitions with each other, as Larsen and Aone did in 1999 [119]: For each cluster from the ground truth partition $\mathscr{P}_{\mathrm{GT}} := \{C_i^g\}_{i \in M_{\mathrm{GT}}}$, an auxiliary partition $\mathscr{P}_i^g := \{C_i^g, X \smallsetminus C_i\}$ is formed. Given a hierarchy $\mathscr{H} := \{P_i\}_{i=1}^h$, all F-scores for the hierarchies partition members are calculated and the highest F-score is chosen. i.e. the F-score for hierarchies is applied in the following way: $\mathrm{FSc}'(\mathscr{H}, \mathscr{P}_{\mathrm{GT}_j}) := \max_{i=1}^h \mathrm{FSc}(\mathscr{P}_i, \mathscr{P}_{\mathrm{GT}_j})$ for all $j \in M_{\mathrm{GT}}$. The overall F-score is then calculated by:

$$\mathrm{FSc}'(\mathscr{H}, \mathscr{P}_{\mathrm{GT}}) := \frac{\sum_{i \in M_{\mathrm{GT}}} (|C_i^g| \cdot \mathrm{FSc}(\mathscr{H}, \mathscr{P}_i^g))}{\sum_{i \in M_{\mathrm{GT}}} |C_i^g|}$$

Figure 5.5 shows a dendrogram with the partitions that are used to calculate the F-score for this hierarchy. While the measure detects different clusterings very well, it also has one apparent problem: It evaluates partitions that do not form part of the hierarchy, i.e. $\mathscr{P}^*$ is the final partition that is compared with the ground truth but $\mathscr{P}^* \notin \mathscr{H}$. This is due to the fact that they do not result from a threshold based partitioning of the hierarchy. I.e. the clusters generally appear on different levels. Looking at Fig. 5.5 the pink cluster belongs to another partition than the blue and the green one. We therefore propose a different approach.

5.7.3 The Partition Score

The easiest solution for the problem of not considering real partitions would be to calculate the F-score $\mathrm{FSc}(\mathscr{P}_i, \mathscr{P}_{\mathrm{GT}})$ for all $\mathscr{P}_i \in \mathscr{H}$ and then chose the maximum of the gained F-scores. A possible problem is that TP, TN, FP, and FN do not take the distance between two clusters into consideration. I.e. in the

5.7 Evaluation of Hierarchical Clustering

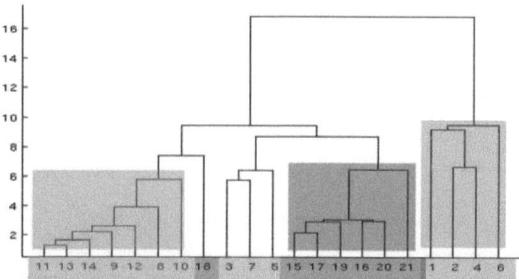

Figure 5.5: classical F-score for hierarchical clustering as defined by Larsen and Aone.

dendrogram in Fig. 5.5 one would count parts of the pink clusters situated in the middle $\{3, 7, 5\}$ as well as $\{1, 2, 4, 6\}$. These two parts of the pink cluster are very far apart from each other and a labeled clustering result would not group them close to each other.

We therefore propose a validation scheme that firstly chooses one of the partitions that form actually part of the hierarchy. Secondly, each clustering label from the ground truth partition is only given to one cluster in the partition that forms the basis of the evaluation. The score of the clustering would then be the maximal possible number of true positives in relation to the total number of samples. Before defining the partition score we have to explain, how we want them to be labeled.

Definition 5.17 For each partition in the hierarchy $\mathscr{H} \ni \mathscr{P}_i := \{C_j^i\}_{j \in M_i}$ and the ground truth partition $\mathscr{P}_{\mathrm{GT}} = \{C_j^g\}_{j \in M_{\mathrm{GT}}}$ we calculate the **partition score** on the space of all partitions $\mathfrak{P}_O$ on a finite set O by

$$\mathrm{PSc}(\mathscr{P}_i, \mathscr{P}_{\mathrm{GT}}) := \max_{\mathrm{L}: M_i \to M_{\mathrm{GT}}^*} S(\mathrm{L}(M_i)) \qquad (5.12)$$

5 Automatic Segmentation of Spectral Data

where $S(\mathrm{L}(M_i)) := \sum_{j \in \mathrm{L}(M_i)} \frac{|C_j^g \cap C_{\mathrm{L}^{-1}(j)}^i|}{|X|}$, and $\mathrm{L}: m \mapsto \mathrm{L}(m)$, injective

with $M_{\mathrm{GT}}^* := \begin{cases} M_{\mathrm{GT}} \cup \overbrace{\{0, ..., 0\}}^{|I|} & \text{for } |I| := |M_i| - |M_{\mathrm{GT}}| \quad \text{if } |M_i| > |M_{\mathrm{GT}}| \\ M_{\mathrm{GT}} & \text{else} \end{cases}$

Calculating the total score for the hierarchy is then defined by

$$\mathrm{PSc}_{\mathscr{H}} = (\mathscr{H}, \mathscr{P}_{\mathrm{GT}}) := \max_{i=1}^{h} \mathrm{PSc}(\mathscr{P}_i, \mathscr{P}_{\mathrm{GT}}) \qquad (5.13)$$

In Equation 5.12 we iterate over all possible labellings of M_i with the ground truth labels from M_{GT}. The labeling function L needs to be injective as every label should only be given once. The codomain can therefore not simply be M_{GT} because, as soon as we have more clusters in $\mathscr{P}_i$ than in $\mathscr{P}_{\mathrm{GT}}$ the injectivity cannot be fulfilled. For this reason, we introduce a set of empty, or "0"- labels $\{0_i\}_{i \in I}$.

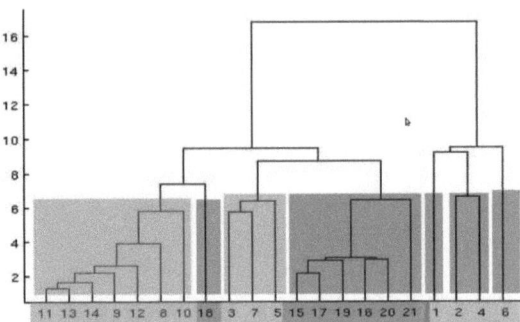

Figure 5.6: Partition Score Calculation.

Fig. 5.6 shows one of the cluster labellings that would be chosen as optimal in the example dendrogram. It becomes clear that all ground truth labels are only given once and that the basis of the labeling is a real partition. This labeling is generally not unique.

5.7.4 Implementing the Partition Score

To find the correct maximum when calculating PSc, one has to consider $\frac{|M_i|!}{(|M_i|-|M_{\mathrm{GT}}|)!}$ (w.l.o.g. that $|M_i| \geq |M_{\mathrm{GT}}|$) possibilities for all $i = \{1,...,h\}$. Depending on the number of ground truth classes and the size of the dendrogram the evaluation can be very inefficient or even not feasible. Therefore, we propose an efficient way to calculate an approximation of the partition score. We start with the coarsest partition (consisting of one cluster, then two clusters a.s.o.) and then use the first optimal partitions as a basis for further labeling. The procedure is described in detail in Algorithm 5.6. It hast to be said, that there generally is no unique optimal labeling but a whole set. The initial labeling in Step 1 satisfies Equation 5.12. Furthermore, the existence of j in Step 2 is ensured because the hierarchical clustering algorithms that are discussed here all work with binary partitioning.

Note that especially with $|M_i| \gg |M_{\mathrm{GT}}|$ this implementation increases the efficiency of the method. It does not depend on the total number of samples. Instead of $\frac{|M_i|!}{(|M_i|-|M_{\mathrm{GT}}|)!}$ we only have to iterate over $(|M_{\mathrm{GT}}| + 1)|M_{\mathrm{GT}}|$ updating possibilities in each step.

5.8 Conclusion

After this chapter we are now able to execute automatic data processing on the THz-TDS data. We found hierarchical clustering to be an adequate method for that purpose due to its independence from previously chosen parameters. The drawback of computational efficiency is compensated by the efficient Chameleon algorithm the functionality of which we explained.

If computational efficiency is not the predominant issue we can still use the classical HC algorithm. It is easy to implement and has a variety of adaption possibilities that can be used. It usually gives stable results and with the linkage functions introduced here, produces no inversions. We can chose either the classical Euclidean distance for the data or the cosine distance that we proposed to be more appropriate for THz-TDS spectra, especially when not baseline corrected.

If computational efficiency is the predominant issue and the classical approach does not yield satisfying results we will use one of the Chameleon based al-

5 Automatic Segmentation of Spectral Data

Algorithm 5.6: Proposed Efficient Partition Score Estimation

Data: Hierarchy $\mathcal{H}$ of data set O and ground truth partition $\mathscr{P}_{GT}$
Result: Partition $\mathscr{P}_F \in \mathcal{H}$ with $\text{PSc}(\mathscr{P}_F, \mathscr{P}_{GT}) \geq \text{PSc}(\mathscr{P}_i, \mathscr{P}_{GT})$ for all $\mathscr{P}_i \in \mathcal{H}$

Initialize $\mathscr{P}_F$ with the coarsest partition: $\mathscr{P}_F := \mathscr{P}_n = \{C_1^n\}$, where $C_1^n = X$

1 Initialize labeling $L_F := L^n : M_n \to M_{GT}$ with $L^n(1) = m$ with $m \in M_{GT}$ and $|C_m^g| \geq |C_i^g|$ for all $i \in M_{GT}$

Initialize $S_F := S(L^n(M_n))$ as defined in Equation 5.17

for $i = n$ **to** 2 **do**

2   Determine $j \in M_i$ for which exist j_1 and $j_2 \in M_{i-1}$ such that $C_j^i = C_{j_1}^{i-1} \cup C_{j_2}^{i-1}$

    We have $M_{i-1} = M_i \smallsetminus \{j\} \cup \{j_1, j_2\}$

    for $k \in M_{i-1} \smallsetminus \{j_1, j_2\}$ **do**
       $L^{i-1}(k) = L^i(k)$

    Initially set $L^{i-1}(j_2) = L^{i-1}(j_1) = 0$

    for $J \in \{j_1, j_2\}$ **do**
       Find best label for cluster J:
       for $m \in M_{GT}$ **do**
            $L_{Tmp}^{i-1}(J) := m$
            Re-assign previous m-label $L^{i-1}((L^i)^{-1}(m))$ either with zero or with one optimal remaining label
            if $S(L_{Tmp}^{i-1}(M_{i-1})) > S(L^{i-1}(M_{i-1}))$ **then**
                $L^{i-1} := L_{Tmp}^{i-1}$

    $\text{PSc}(\mathscr{P}_{i-1}, \mathscr{P}_{GT}) = S(L^{i-1}(M_{i-1}))$
    if $PSc(\mathscr{P}_{i-1}, \mathscr{P}_{GT}) > S_F$ **then**
       $\mathscr{P}_F := \mathscr{P}^{i-1}$
       $S_F := S(L^{i-1}(M_{i-1}))$

5.8 Conclusion

gorithms. The classical Chameleon algorithm's similarity measure was slightly altered to fit to the useful definitions that were made beforehand. Alternatively we proposed to use Phase I of Chameleon together with the classical hierarchical clustering to be able to use the different distance measures that are supplied for this algorithm. We furthermore discussed two approaches on including spatial similarities. Algorithms were given and may be used to improve a segmentation of a hyperspectral image.

In order to evaluate the clustering result we proposed a new evaluation scheme for hierarchical clustering. We felt the necessity for a new scheme because the classical approach evaluates hierarchies that were not part of our solution space.

5 Automatic Segmentation of Spectral Data

6 Application Examples

It was described in the introduction that the visualization of THz-TDS data as well as the extraction of content is difficult. Methods needed to be developed for this purpose. The different techniques that were proposed during the course of this thesis have the objective of enabling the content extraction from THz-TDS data in an efficient manner. To show the usefulness of the developed algorithms and approaches, we will now present some application examples. All of these examples were not analyzed beforehand but were rather presented to us in the initially measured form. Hence, transforming this raw hyperspectral data into an interpretable form with respect to different application scenarios was a strong motivation for this thesis. On the one hand the goal of this chapter is to show that the methods proposed and discussed throughout the last chapters can be applied on different kinds of measurements. On the other hand we want to illustrate in what context they are applicable and what should be considered when using them. Application scenarios can be within the analysis of single spectral measurements as well as hyperspectral images. Although all presented measurements are actually acquired hyperspectrally, we consider both application cases.

Contribution *All data processing shown in this chapter is a contribution of this thesis. To link the presented data sets to the respective applied methods, we will give a tabular overview of the measurement and the sections or the chapter in which these methods were discussed.*

Section in this chapter	Used methods and section were they were introduced
Section 6.1: Eight THz-TDS images of chemical compounds	Preprocessing: Section 2.3 Wavelet based features: Section 4.2 Advantages of cosine distance: Section 5.3 Evaluation with partition score: Section 5.7

113

6 Application Examples

Section in this chapter	Used methods and section were they were introduced
Section 6.2: THz-TDS image of pressed lactose pellet	Echo removal: Section 2.2
Section 6.3: THz-TDS image of mockup mail bomb	PFS features: Excursus 2 based on previous section Simplified Chameleon: Section 5.5 compared with classical version Including spatial characteristics: Section 5.6
Section 6.4: Raman image of polymer foil	Applicability to other spectral domain: Wavelet features: Section 4.2 Chameleon based clusterings: Section 5.5

6.1 Eight Chemical Compounds

The first data set was acquired by the Fraunhofer Institute for Physical Measurement Techniques (Fraunhofer IPM) in Kaiserslautern and consists of eight hyperspectral measurements of different chemical compounds, with 5 × 5 pixels. Each signal is composed of 4268 channels, sampled at a rate of 0.0312 picoseconds. The chemical compounds are pressed into pellets as displayed in Figure 6.1. The compounds are p-amino benzoic acid (PABA), acetyl salicylic acid (ASS), glucose, saccharose, lactose and tartaric acid. There are two pellets that contain glucose — one with a high concentration of the agent one with a lower concentration. In all of these measurements, not the full pellet is acquired but only a square taken from the middle. Each pixel covers an area of 1 mm^2. Therefore, the measurements are not typical hyperspectral imaging measurements with unknown content and various objects. On the contrary, each image contains one pure chemical.

We will use the images for two purposes. Firstly, we want to illustrate the adequacy and necessity of the various proposed preprocessing methods in applications where the characterization of chemicals is needed. Secondly, we will evaluate the feature selection and the positive effect of the cosine distance by using the proposed cluster evaluation technique. We will do this by creating an artificial hyperspectral image put together from the single chemical measurements and thereby gain a ground truth that is necessary for the evaluation.

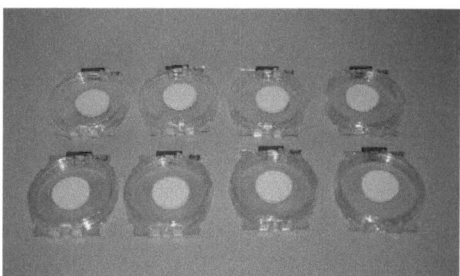

Figure 6.1: Pressed pellets with chemical agents.

6 Application Examples

6.1.1 Comparability of Chemical Measurements

Other spectroscopic techniques that are used to characterize chemicals, such as infrared spectroscopy, have databases that contain the spectral expression of many compounds [17]. Unfortunately this can not be said for THz spectra. As an emerging technique, not as many different compounds are measured yet and even if they are measured the standardization of the measurements is low. The existing databases, such as [121], declaredly differ in quality and method of acquisition. For many applications this is a problem because they aim at detecting specific compounds. Having a reliable database that contains peak positions and relative intensities for each compound is therefore necessary. For this purpose we propose to first use a fixed pipeline of preprocessing steps to make the spectra comparable. Afterward a procedure to extract each compounds peak positions and depth can be applied.

Methods

The preprocessing pipeline we propose is described in Algorithm 6.1. It is based on various algorithms introduced in Chapter 2. In addition to the standard

---
Algorithm 6.1: Determine Peak Position and Relative Height for One Compound

Data: THz-TDS measurement in time-domain
Result: Frequency-domain transmittance with constant baseline
1 Remove echo pulses as detailed in Algorithm 2.2
2 Apply Fourier transform and divide by reference as in Section 2.1
3 Determine dynamic range (DR) as in Algorithm 2.3
4 Apply baseline correction as in Algorithm 2.4

---

procedures (Step 2) we apply the proposed echo removal (Step 1) in the time-domain beforehand. in the frequency domain we then analyze the DR of each measurement in Step 3 and then estimate a baseline and correct the spectra in Step 4.

After preprocessing the data such that the spectra are now generally comparable, in Algorithm 6.2 we propose a procedure that can be used to extract

6.1 Eight Chemical Compounds

each compounds peak position and depth if various samples per compound are available. At the peak candidate detection of Step 1 of Algorithm 6.2, min-

---

Algorithm 6.2: Find Peak Position and Depth for one Compound

Data: Multiple THz-TDS spectra $\{S_i\}_{i=1}^{n}$ of one compound X with constant baseline, limited to DR of the frequencies $\Omega := \{\omega_1, ..., \omega_{\mathrm{maxDR}}\}$ Where $maxDR$ is the maximal cutoff point of S_i

Result: Positions $\Omega*$ and relative depth $D : \Omega* \to [0, 1]$ of peaks of compound X

1 **for** $i = 1$ **to** n **do**
   Find all frequencies $\Omega_i := \{\omega_{i_1}, ... \omega_{i_{n_i}}\}$ that contain minima of S_i

2 Build histogram $H : \Omega \to \mathbb{N}$ with $\omega_i \mapsto |\{i \in \{1, ..., n\} | \omega_i \in \Omega_i\}|$
   for $i = 1$ **to** $maxDR$ **do**

3   **if** $H(\omega_i) \geq \frac{|I|}{3}$ **then**
         Add frequency position: $\omega_i \in \Omega*$
         Add mean value in all measurements: $D(\omega_i) = \sum\limits_{j=1}^{n} \frac{S_j(\omega_i)}{|I|}$

---

ima can potentially be found that do not originate from an absorption peak but rather only noisy oscillations in the spectrum. Especially if the concentration of the agent is low, considerably more peak positions are found than if it is high. Therefore, it is important to have a histogram over several measurements and their peak position candidates. Building this histogram, however, is the most sensitive step of Algorithm 6.2. The default bin size is one bin per available frequency channel $\omega \in \Omega$. However, in higher frequent regions (i.e. the regions with worse signal to noise ratio) the peak positions tend to vary slightly between measurements. Therefore, we decided to use a frequency dependent bin size. From 0.1–1 THz one frequency channel per bin, from 1–2 THz two and for frequencies higher than 2 THz we use four. We then regard a minimal point to be a peak if the histogram counts $\frac{1}{3}$ of the spectra to contain it in Step 3. This threshold is set low because no peak should be missed. Generally the downside of such a low threshold is a misclassification of non-peaks.

6 Application Examples

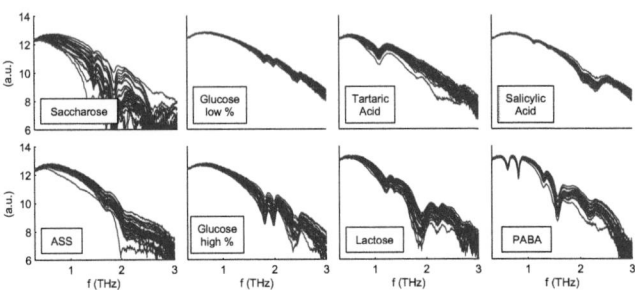

Figure 6.2: Original magnitude spectra after time-domain preprocessing.

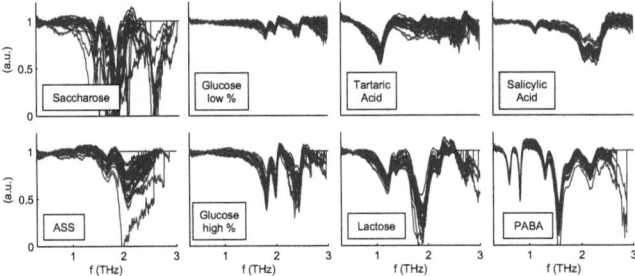

Figure 6.3: Baseline corrected spectra from Figure 6.2. Information beyond the DR is discarded.

Results

We now show the applicability of the algorithms to the chemical compound images. For this purpose we treat them as if they were 200 independently acquired spectra. Nevertheless, they are visually grouped together (per compound) to show that the methods increase the similarity among the spectra. Figure 6.2 shows the original spectra, i.e. 25 per compound with the typical magnitude spectra that can be found in THz-TDS measurements. Depending on the measurement, the depth of the peaks and the variance differs and all have a clearly non-constant baseline. Figure 6.3 displays the result of the preprocessing steps from Algorithm 4. The spectra now have a constant baseline and are only con-

6.1 Eight Chemical Compounds

sidered until the end of their DR. Everything beyond the DR is set constant and therefore introduces no false distances or similarities. Note that the saccharose measurements have a high variance, therefore, some of the spectra get cut-off quite early (at about 2 THz). The other spectra do not share this problem.

For this pipeline the only parameters that are set, are a minimal and a maximal possible DR. Set here between 2 and 4 THz, the algorithm is not sensitive to them. They could be neglected altogether but high variances in the earlier regions of the spectra can lead to an unwanted cut, especially when the intensity of the compound is low, as for example in the low concentration glucose measurements.

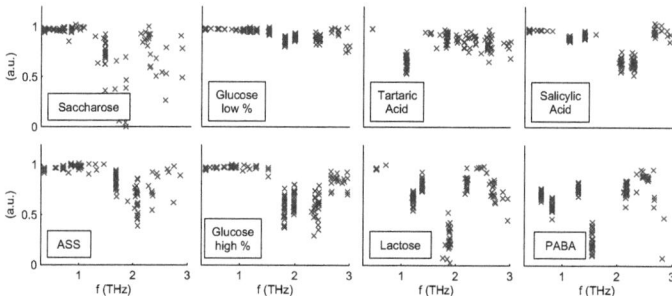

Figure 6.4: Positions and values of all detected minima of compounds from Figure 6.3.

After the preprocessing, we applied Algorithm 6.2. The peaks that are detected on the 7 compound data are shown in Figure 6.6. The result of the peak candidate detection of Step 1 of Algorithm 6.2 is shown in Figure 6.4. Note that in the low percentage glucose measurement more minima are found than in the high percentage which is not consistent with the fact that it is the same compound. This illustrates the necessity of acquiring several measurements. The histograms of the example data sets are shown in Figure 6.5. To illustrate which peaks are valid, the histograms in Figure 6.5 display their quantities only until they reach the threshold of 8 spectra/bin with a total of 25 acquired spectra per compound.

6 Application Examples

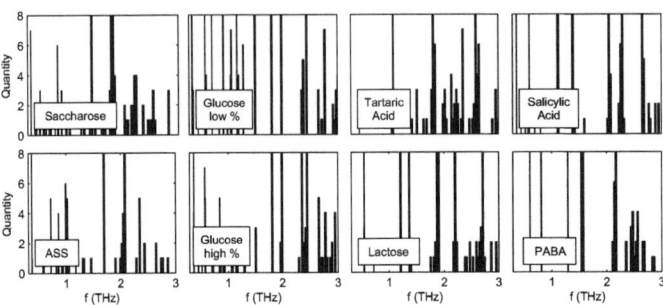

Figure 6.5: Histograms over peak positions of Figure 6.4.

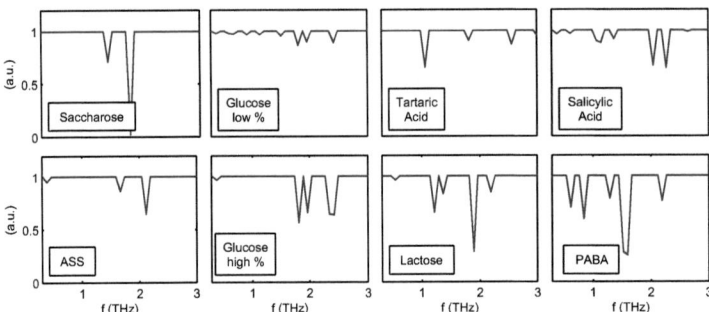

Figure 6.6: Final peak positions for compounds from Figure 6.3.

The automatically detected peaks are consistent with the visual examination of the original compounds. Moreover, even small peaks and slight movements of original transmittance spectra are clearly detected. In several of the measurements, low peaks in the beginning of the spectrum are found. It can not be said if all of these peaks are really based on a characteristic absorption line. In the case of glucose, as we have two measurements with different concentration of the

6.1 Eight Chemical Compounds

agent, it can be said that the peaks in regions between 0.5 and 1.5 THz in the low concentration sample must be noise, otherwise they would also appear in the high concentration measurement. On the other hand, the low 0.5 peak in Lactose is not noise as it is reported in literature [122].

6.1.2 Parameter Choice in Classification

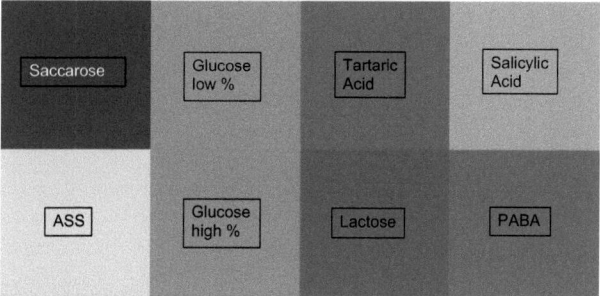

Figure 6.7: Schematic view of the concatenation of the 8 hyperspectral chemical compound measurements.

We have shown in the previous section that it is useful to apply various preprocessing steps to be able to find peak positions and depth and thereby build databases of chemical compounds. The focus of this thesis, however, is on the automatic segmentation of hyperspectral images. In the last chapters we introduced different feature sets that can be used as a clustering basis as well as different parameters for classical hierarchical clustering. Therefore, in this section we want to show the following things:

(1) The ability of the representations of the magnitude spectrum from Section 4.2 to lead to a good clustering result. We will compare the real wavelet

6 Application Examples

base features from Section 4.2.2 with the complex wavelet magnitude and phase from Section 4.2.3 and the full spectra as displayed in Figure 6.2.

(2) The applicability of the classical hierarchical clustering as explained in Algorithm 5.1 and especially the influence the different parameters, particularly the proposed cosine distance, have on the outcome of the clustering.

(3) The possibility of evaluating the hierarchical clustering with the partition score introduced in Algorithm 5.6.

Methods

To build a hyperspectral image with a known ground truth, the 8 measurements of chemicals are used again. We concatenate them into one artificial hyperspectral image. A schematic view of this image, and thereby a ground truth labeling of the image, can be seen in Figure 6.7. The complete image consists of 10 × 20 pixels.

Often when not single chemical compounds but hyperspectral measurements with different content — and also different content quality — are acquired, some of the preprocessing steps are error prone: We have mentioned in Chapter 4 that although the baseline correction works well on measurements with a consistently high DR, it might fail when the measurement quality is worse. Luckily, the proposed PFS coefficients, in addition to reducing the features, perform a zero normalization, which is why they can be applied without prior baseline correction. Hence, we will use the spectra in their original form, i.e. after the standard preprocessing as shown in Figure 6.2. For comparability reasons we furthermore have to use a common DR for all spectra. On the one hand including noisy channels into the clustering can bias the result, on the other hand, important information of other measurements might lay beyond an early DR of one measurement. Therefore, we will use two different DRs as is shown in Figure 6.8.

On the left hand side of Figure 6.8, some of the spectra are exemplary shown. They consist of 512 channels. The right hand side of the figure displays the 16 real wavelet coefficients of the 4th downsampling level representing the same spectral range. Some of them contain characteristics after 350 channels, others approach their noise floor at 300 already. To determine how much the noise on the one hand and the lack of information on the other hand influence the clustering, we compare the cyan colored frames in Figure 6.8 with the the black frame channels.

6.1 Eight Chemical Compounds

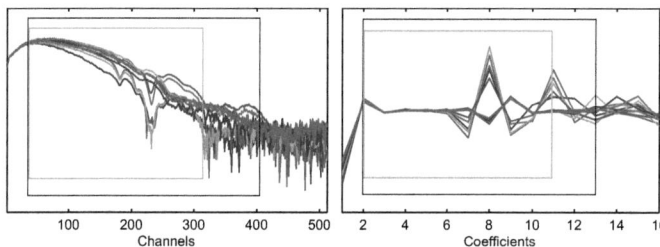

Figure 6.8: Channels that are used for clustering are framed. Big data set is black, small data set is cyan. Left: Full spectra; Right: Real wavelet coefficients.

We will furthermore compare 6 different feature sets for the hierarchical clustering and denote them as follows:

(1) Black, Real. Real Wavelet coefficients channels used are PFS 8–19, i.e. 12 channels

(2) Black, Full. Full spectrum with channels 32–416.

(3) Black, C-Phase. Phase of complex wavelet coefficients PFS 8–19.

(4) Black, C-Mag. Magnitude of complex wavelet coefficients PFS 8–19.

(5–8) Cyan Real, Full, C-Phase and C-Mag, i.e. PFS 8–17, i.e. 10 channels, representing only the original channels 32–352.

Furthermore we use 6 sets of parameters.

(1) Euclidean. Euclidean distance between samples

  (a) Complete. Distance between clusters is maximal distance between samples

  (b) Single. Distance between clusters is minimal distance between samples

  (c) Average. Distance between clusters is average distance between samples

(2) Cosine. Cosine distance between samples with 3 linkage functions

We use the partition Score PSc to compare the clustering results with the different parameters on the basis of the feature sets with the ground truth partition.

123

6 Application Examples

Results

		Euclidean			Cosine		
		Complete	Single	Average	Complete	Single	Average
Black	Full	64.5% (17)	56.5% (85)	64.5% (30)	61% (33)	51.5% (100)	59% (45)
	Real	67.5% (42)	71% (59)	73% (48)	94% (10)	84.5% (12)	96% (12)
	C-Phase	69.5% (11)	73% (25)	73% (20)	64.5% (13)	75.5% (18)	73.5% (15)
	C-Mag	63% (22)	56% (57)	64% (33)	83.5% (15)	80% (44)	88.5% (15)
Cyan	Full	61.5% (15)	82.5% (42)	73% (27)	63% (28)	77.5% (46)	67% (35)
	Real	74.5% (29)	84% (39)	84% (32)	98.5% (9)	87% (9)	98.5% (9)
	C-Phase	70% (18)	74% (27)	71% (22)	74.5% (14)	74.5% (24)	72% (20)
	C-Mag	60.5% (38)	69.5% (40)	69.5% (30)	90.5% (13)	90% (28)	91% (14)

Table 6.1: Partition Scores as introduced in Algorithm 5.6 for different feature sets and distance-linkage combinations together with respective optimal clustering level (in braces below the score).

Two kinds of results are displayed in Table 6.1 the best scores and the level at which they are achieved. The scores are named by the percentage of correctly clustered samples in relation to all labeled ground truth samples. Below this percentage the level at which this result is achieved is given, also. The higher the score, the better the result and furthermore the lower the level, the easier it is to detect this "optimal" result. In each row the optimal score is colored red.

Regarding the distance function, the cosine distance performs better than the Euclidean distance. Most of the optimal scores are yielded by the cosine distance. Although the optimal results for the full spectral clustering are gained with an

6.1 Eight Chemical Compounds

Euclidean distance they do not differ much from the results achieved with the cosine distance. In contrast to that, the cosine distance very much improves the clustering result for the wavelet based feature sets — particularly the complex magnitude and the real wavelets.

As to the linkage function, it can not be concluded that one clearly outperforms the other. The average linkage achieves the highest score in most cases but the other linkage functions are not far from it. Single linkage performs mostly worse than the other two, however, especially for the wavelet based features.

The cyan feature sets perform better than the black ones, which means that leaving out information leads to less problems than including too much noise. Furthermore the full spectra produce the worst clustering results, with a score of around 60% at best for the black region whereas the real wavelets achieve a 90% score. The phase of the complex wavelets is least influenced by the width of the chosen coefficients, but otherwise performs worse than the other two small feature sets. The complex magnitude gives comparable but slightly worse results as the real wavelets

In addition to the displayed table we show the optimal segmentation for the cyan feature sets in Figure 6.10a — 6.10d. The label colors are the same as in the ground truth image in Figure 6.7, the zero-label as defined in Algorithm 5.6 is displayed in black. The compounds that are generally best detected are the PABA (lilac) and the glucose (green) with the high concentration. The compound that is clustered worst is the saccharose. This result complies with the quality of the spectra (see Figure 6.2). The complex phase is very sensitive to shifts and interferences, it therefore gives different misclassifications than the other ones. The glucose with high and low absorption tends to be clustered together, especially when using the cosine distance. As the cosine distance applies a normalization on the peaks this is not surprising but this normalization should be kept in mind when the distance is applied. It should not be used if the concentration rather than the peak position is the discriminating characteristic.

6 Application Examples

Figure 6.9: Best segmentation of complex wavelet magnitude feature set (small) for the different parameter combinations.

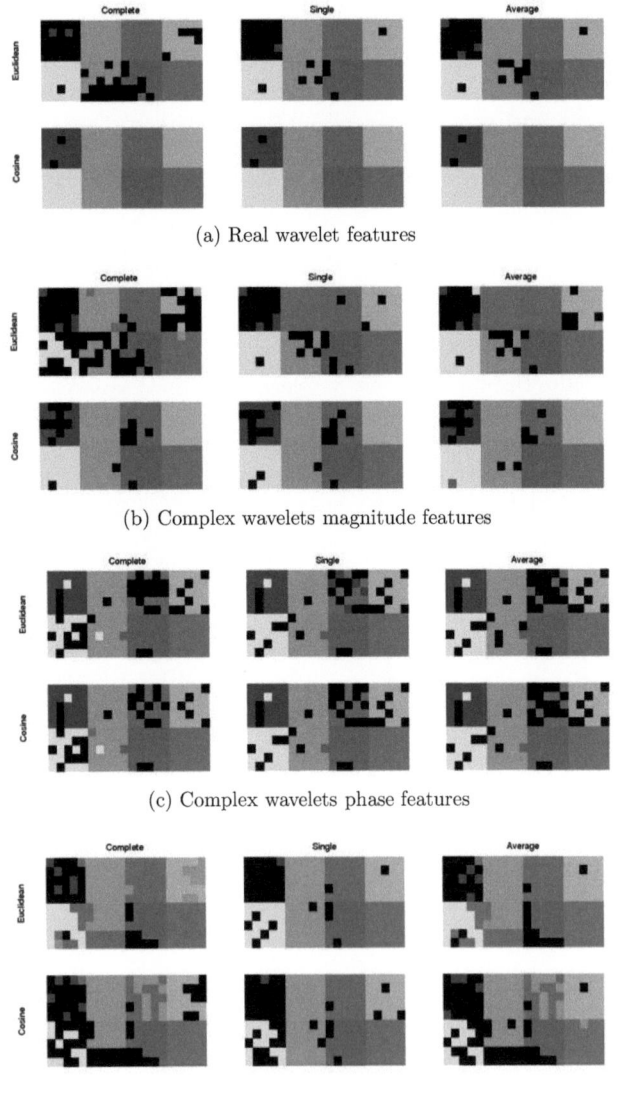

(a) Real wavelet features

(b) Complex wavelets magnitude features

(c) Complex wavelets phase features

(d) Full Spectrum

6.2 Lactose Pellet

The next data set was acquired by K.Wiesauer and S.Katletz from RECENDT GmbH, Linz. It is a hyperspectral measurement of a pressed lactose pellet with 32 × 32 pixels. Each signal is composed of 512 channels sampled at a rate of 0.1334 picoseconds. The image consists of a non-absorbing background and the pellet in the lower middle. The measurement is not difficult to segment, as the lactose pellet is not concealed and lactose with this polymorphous shape is easy to identify due to a clear peak at 0.5 THz. We want to illustrate two things on this measurement: Firstly the necessity of automatic echo removal in hyperspectral imaging and secondly differences between the feature sets.

Methods

For echo removal we use the wavelet shrinkage based method introduced in Section 2.2. As a feature representation, the PFS features from Excursus 2 are then applied. The first six features illustrate the difference between windowed and unwindowed spectra. Furthermore we focus on the differences between the three representations of the magnitude spectrum. Those are the integration over a fixed set of intervals on the one hand and the two methods we proposed, i.e. PFS features 7–X based on real-wavelets from Section 4.2.2 and based on complex wavelets from Section 4.2.3 on the other hand.

Results

The measurement is particularly interesting because it has a strong reflection pulse of the reference at the positions where the pellet is located. An unpreprocessed example signal from a pellet pixel is shown by the red spectra on the left hand side of Figure 6.10. The blue signal in the same figure is the result of the echo removal. Although a big part of the signal (almost 10 picoseconds) has to be removed the main peak is preserved. The right hand side of Figure 6.10 shows the shape, the windowed and unwindowed signals have in their magnitude spectra. Instead of many interferences in the red unwindowed spectrum, the blue spectrum depicts one clear and broad peak at the right position.

It was said already that the image contains only two types of spectra, namely the background and the said lactose spectra. We show the background spectra on the right hand side of Figure 6.10 in green. They are mostly smooth over the whole DR and are not changed by the echo removal. It has to be remarked

6 Application Examples

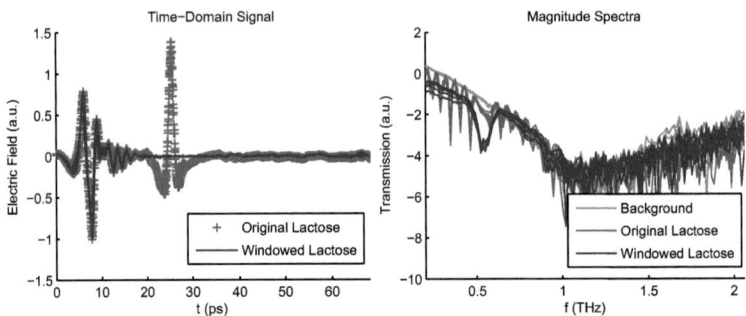

Figure 6.10: Lactose signals of hyperspectral measurement and respective magnitude spectra including spectra of the background signal with and without echo removal.

that the proposed echo removal does not need any parameters. It can hence be applied on a whole image without adaption to the different kinds of spectra. It will not lead to information loss if no echoes are detected.

We now display the PFS representations of the images. The first six features for both kinds of signals are shown in Figures 6.11 and 6.12. In case of the unwindowed spectra, the main amplitude which the first two features are based on is the highest one. The highest pulse in the lactose spectra is the "wrong" one, namely the reference remainder. Its position coincides with the positing of the background pixel main pulse, therefore the set-off in the second PFS feature is non-existing in the unwindowed version. In the windowed case on the other hand, there are no echoes anymore, because the windowing produces echo-free signals. Hence, in case of Figure 6.12, the 5th features is a constant image. Generally the windowed features show the Lactose pellet and especially in the less discrete features, namely the wavelet scaling factor (6) and the amplitude height(1), the windowed spectra give more clear results.

Another aspect we want to emphasize with this data is the difference between the representations of the magnitude when analyzing only chemical content. Regarding the windowed lactose pellet, the background only consists in one dominant peak, therefore in case of the complex wavelets we will only use the magnitude and not the phase. Due to the downsampling scheme of the wavelet transform, for this measurement, each of the coefficients covers ~ 110 GHz. Hence,

6.2 Lactose Pellet

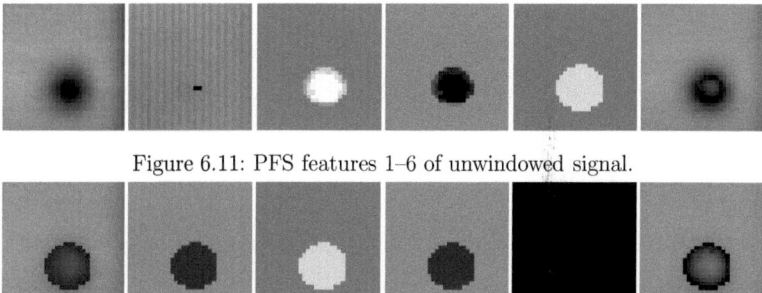

Figure 6.11: PFS features 1–6 of unwindowed signal.

Figure 6.12: PFS features 1–6 of windowed signal.

Figure 6.13 shows the respective integration over 110 Ghz intervals covering the DR which is about 1 THz. Figure 6.14 shows the real wavelet coefficients based on the Battle-Lemarie wavelet and Figure 6.15 shows the magnitude of the complex wavelets.

All these feature sets are sufficiently discriminating for a segmentation of this image. Nevertheless the integration over the full spectra shows more or less the same difference in all intervals. This is due to the dominance of the overall absorption. In contrast to that, because of their inherent normalization to zero, the real wavelet coefficients as well as the complex ones show the main differences in the regions around 0.5 THz in the 3rd and 4th coefficients. Comparing the real and the complex wavelet coefficients the differences are slightly finer. It was already mentioned that the real wavelet transform has the disadvantage of being non-directional and non-shift invariant. Therefore, in Figure 6.14 in interval 0.42–0.53, the lactose pellet is smaller (darker) than the background while in the next interval it is higher (brighter). The original peak is of course only one-directional. This is better represented by the complex magnitude. On the other hand the complex coefficients seem to be more sensitive to noise in non-peak regions, such as the last two coefficients.

6 Application Examples

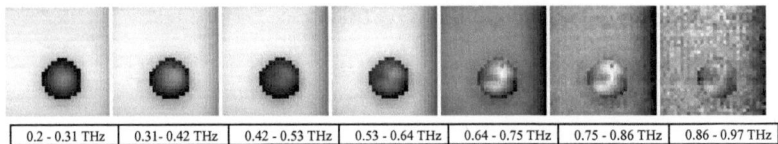

Figure 6.13: Integration of full spectra in each named interval.

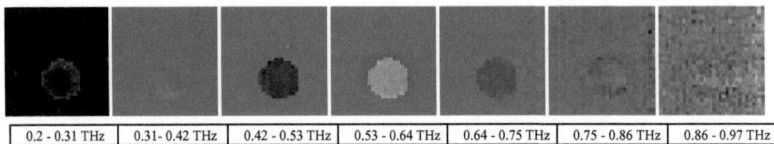

Figure 6.14: Real Wavelet coefficients.

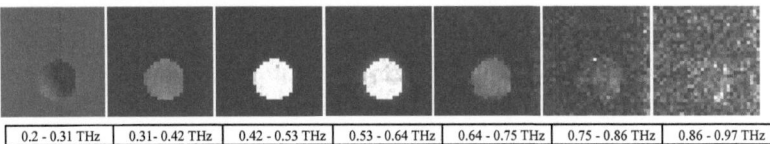

Figure 6.15: Complex wavelet magnitude coefficients.

6.3 Mockup Mail Bomb

The last THz measurement was acquired by Daniel Molter of the Fraunhofer IPM in Kaiserslautern. It is a hyperspectral image of a mockup mail bomb with 99 × 198 pixels. Each signal has 2600 channels, acquired at a sampling rate of 0.02 picoseconds. The content consists of a closed envelope containing different types of material. The envelope itself is shown on the left hand side of Figure 6.16 while the right hand side displays its content. The dotted line marks the area that is measured. The content consists of highly absorbing components such as a micro-controller and metallic wires, plastic content as well as two satchels containing chemical powder (Lactose and Salicylic Acid). The application area of such a measurement is non-destructive mail testing. In this application one wants to detect the metallic compounds on the one hand. Detecting metallic components is possible with other techniques such as x-ray imaging, however. Therefore, the special focus must be on the detection of the satchels with chemical content on the other hand. We will use this measurement to illustrate that various techniques introduced in this thesis can be used to perform such a detection and to demonstrate how they can be used. The focus will be on two things:

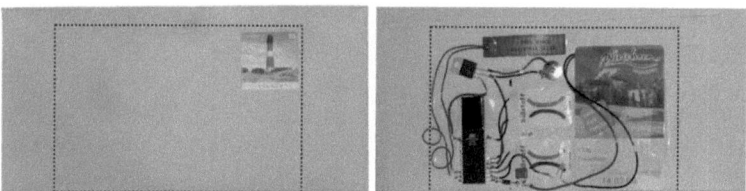

Figure 6.16: Content of mockup mail bomb envelope with wired compounds, a micro-controller, and satchels containing chemicals.

(1) The usefulness of the proposed PFS coefficients. We will show all of these features for this measurement to illustrate the kind of content that can be extracted from them.

6 Application Examples

(2) The usefulness and necessity to apply the Chameleon based clustering algorithms and also the improvement that can be achieved by including spatial features. Both Algorithm 1 and Algorithm 5.5 will be used.

6.3.1 PFS Illustration

We have already depicted some differences between the usable features in the previous section. However, within the mockup mail bomb there is a larger variety of content than in the lactose pellet measurement. Not only chemicals can be found but also plastic and metallic material. Furthermore, the measurement is acquired through a closed envelope — as shown in Figure 6.16. We therefore use this measurement to illustrate how the PFS coefficients can be used to identify and differentiate different material in general and especially to split the chemical peak position information from the overall absorption.

Methods

Similar to the previous section we use the PFS features from Excursus 2. The first five features can be used to identify the different materials within the envelope. The absorption and time delay of the main amplitude pulse — PFS 1 and 2 — are often applied by physicists to illustrate the validity of the measurement. Especially PFS 1 gives a first impression of the measured content. Various kinds of absorbing materials can be outlined by it already. If no absorbing material would be present a constant image would be visible. The time delay implies information about the refraction of the objects. The phase slope and intercept show similar characteristics. The echo pulse position gives information about reflection within the material, which is especially useful for multilayered plastics. The wavelet scaling factor (PFS 6) is a feature that includes the overall absorption information contained in the magnitude. We will show both the complex and the real scaling factor to illustrate how well they separate the overall absorption from the peaks.

Although various content can be made visible in the time-domain and by simple phase features, the focus in this measurement lies on the chemicals. Therefore, we display information gained from the magnitude spectrum. The goal is to be able to extract the absorption peaks and use them to segment the chemical satchels. To illustrate the improvement, the proposed representation presents over the classical methods, we begin by showing how the magnitude spectrum is

6.3 Mockup Mail Bomb

often represented.

Similar to the last application example, the integral over the region of interest, i.e. the DR is taken piece-wise. Instead of looking at each channel, one can thereby subsume information from a whole interval. To compare this method to our proposed method, we will divide the DR (we consider 0.3 to 1.7 THz) into as many intervals as features are used for the wavelet coefficients and integrate the magnitude over these intervals. We will then compare this to the complex and real PFS features 7–X. One of the special characteristics of this representation is that one can bring it into direct relation to the original spectra, i.e. one can determine positions of peaks on the basis of the representations. To illustrate this, we will plot the features directly above the spectral regions they are representing.

Results

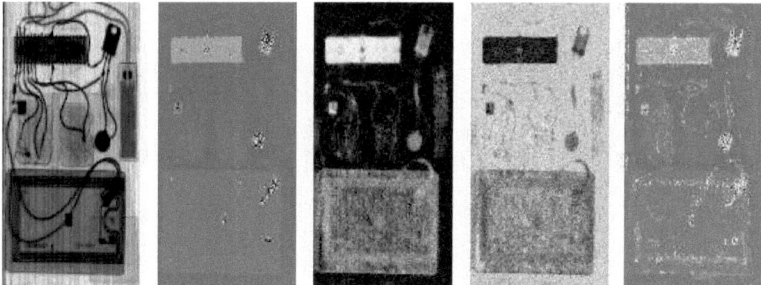

Figure 6.17: Left to Right: Time delay of the main amplitude (PFS 2), the slope and intercept of the phase spectrum (PFS 3 and 4), and echo pulse positions (PFS 5).

The first five PFS coefficients are shown in Figure 6.17. The two images on the left hand side are the absorption and time delay of the main amplitude pulse. Especially PFS 1 outlines the different material, while PFS 2–4 show characteristics about the refraction. For example, while the micro-controller on top and

6 Application Examples

the ski-card at the bottom seem similar in PFS 1, they reveal different refractive characteristics and can thereby be separated. As there are no multilayered materials in this image, the informational content of PFS 5 is low.

Although the satchels containing the chemicals can be outlined in PFS 1, they are only vaguely seen and cannot be distinguished from each other. The absorption peak of Salicylic Acid that can be found within these magnitude spectra, due to the short DR, lie at ~ 1.1 THz and the big peaks of Lactose are at 0.5 and 1.4 THz. Note that the Lactose has a different polymorphous shape than the one used for the peak extraction in Section 6.1, therefore the 0.5 THz peak is more pronounced here.

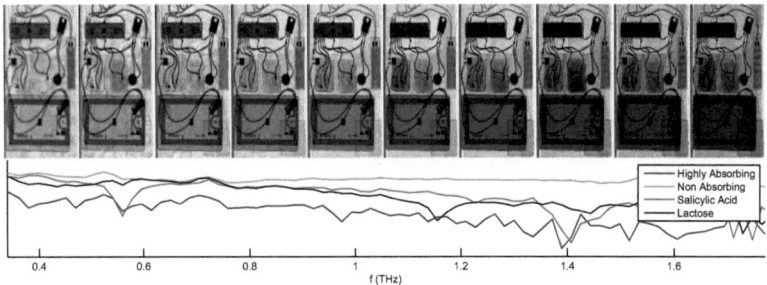

Figure 6.18: Classical view of the magnitude integration representation (top) and the respective intervals that are represented (bottom).

The typical integration information is shown in Figure 6.18. On the bottom of this figure, we plotted example spectra of qualitatively different pixel regions with high absorption, low absorption and of the two chemical compounds. Although the general absorption and the material can be seen in every interval representation, this information unfortunately dominates the images. The absorption peaks of the chemicals should be visible in interval 2, 6, and 7 but are merely visibly as slightly darker shadows.

The proposed feature set has an inherent normalization to zero and thereby overcomes this problem. The absorption is concentrated in PFS 6. For the

6.3 Mockup Mail Bomb

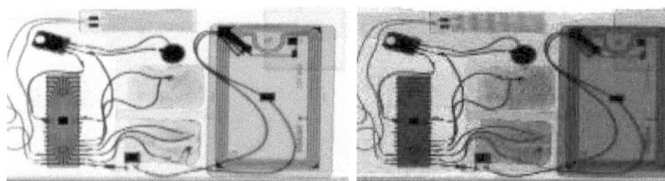

Figure 6.19: Scaling factor of real wavelet coefficients (left) and complex wavelet coefficients (right).

complex wavelet transform's magnitude and the real wavelet transform these are shown in Figure 6.19. The other features PFS 7–X should then contain only information about singularities, i.e. peaks, from an otherwise constant spectrum. Figure 6.20 shows the respective coefficients representing the same region as in Figure 6.18 for the complex and the real case. On the one hand these feature seem to contain less information than the respective integration in Figure 6.18. On the other hand the chemical information can be extracted very easily here. The Peaks at the corresponding positions do not only distinguish the satchels from the rest of the content but also from each other.

Comparing the complex and the real PFS, the real coefficients in the top row of Figure 6.20 separate overall absorption and peaks very well, while the complex coefficients on the bottom contain more information about the overall absorption. However, the advantage of the latter is that the absorption peaks are all pointed to the same direction, i.e. brightness equals absorption. This can not be said for the real coefficients, peaks can show both brighter and darker outcome than the surrounding pixels. This is particularly evident at the wires in the ski-card. In the last three images on top these regions contain very bright and very dark spots, while in the same regions at the bottom they are only bright. Hence, this kind of noise gets increased by the non-directionality of the real wavelet transform.

Figures 6.17, 6.19, and 6.20 shown in comparison to Figure 6.18, point out the advantages of the proposed feature sets and what each feature signifies. Furthermore, the difference between the real wavelet PFS 7–X and complex magnitude wavelet PFS 7–X is illustrated. We will now continue by using some of the illustrated characteristics and apply cluster analysis for an automatic segmentation of the images.

6 Application Examples

Figure 6.20: Real wavelet coefficients (top), complex magnitude coefficients (middle) and respective intervals they represent (bottom).

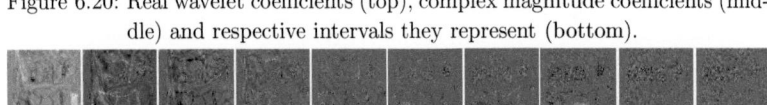

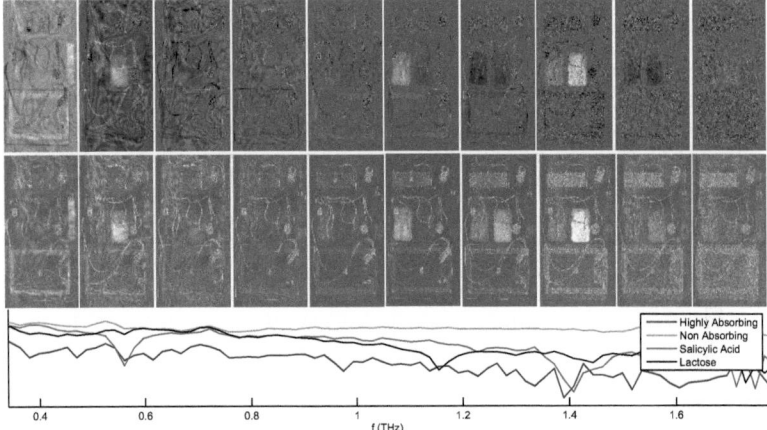

6.3.2 Hierarchical Chameleon Based Clustering

After finding a good feature representation one wants to segment the image with respect to these features. The goal is to find a good segmentation mechanism. During this work, various hierarchical clustering methods were discussed. We will compare those two that are able to handle high-volume images like this one.

Methods

The data volume of about 20000 spectra is not exceedingly high, but already poses problems when processed by the classical hierarchical clustering. Therefore, we use the classical Chameleon from Algorithm 5.2 and the proposed simplified Chameleon from Algorithm 5.3. As one can see from Figure 6.20, the complex magnitude and the real wavelet coefficients show similar results. This is also the case when they are used for clustering. Therefore, we will only show the results for the real PFS features. The range that is depicted in Figure 6.20 will be used, hence, we have 16 features. As we do not just get one segmentation by using hierarchical clustering but a whole hierarchy of segmentations, we do accordingly

6.3 Mockup Mail Bomb

not only display one of them but various cluster levels. We will always display 5 levels. The levels are chosen manually at points where information of interest appears.

Results

Figure 6.21 depicts some cluster levels resulting from the classical Chameleon, while the segmentations from Figure 6.22 result from the simplified version. In

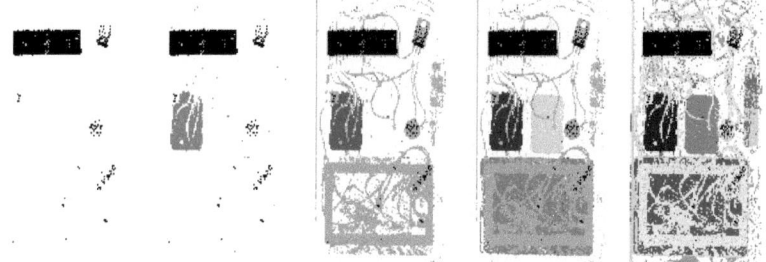

Figure 6.21: Classical Chameleon, cluster levels $1, 2, 3, 6$, and 10.

both cases we chose an initial number of clusters of 500 and built the nearest neighbor graph with 100 neighbors per node. In the classical Chameleon we set $\alpha = 1$ and in the simplified Chameleon we chose to use the Euclidean distance between the centers of the initial partition as the initial distance matrix. Further, an average link clustering was performed.

The result of the classical Chameleon in Figure 6.21 shows that all important objects are detected in early level. Although the same kind of information is detected by the simplified Chameleon, the levels are considerably higher, namely level 10 in comparison to level 19. However, the classical Chameleon is more sensitive to the high noise level in some features. The early wavelet coefficients (PFS 7 and 8) contain wavelike interferences, probably from the structure of the envelope itself, that are suppressed by the simplified Chameleon and clearly visible in early stages of the classical Chameleon. These structures are undesired in the application scenario of automatic mail control, but might be of interest in

6 Application Examples

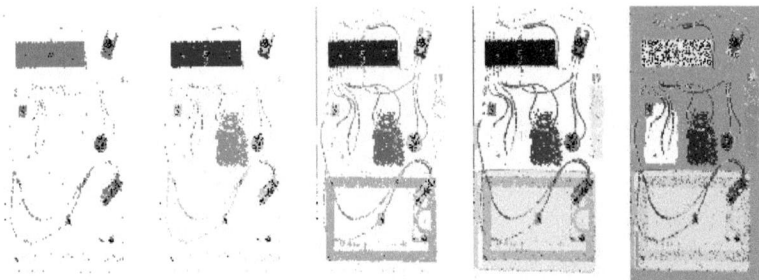

Figure 6.22: Simplified Chameleon with average linkage and Euclidean distance, cluster levels $2, 5, 7, 12$, and 19.

other scenarios. It can therefore not generally be said which algorithm performs better here.

It can be said, however, that a lot of noise is still present in the clustered images, in the next section we will therefore apply the proposed algorithms to include spatial features into a segmentation to improve the overall smoothness.

6.3.3 Spatial Characteristics and Segmentation

Within Figure 6.21 and 6.22 the segementations revealed the information of interest but were still quite noisy. Therefore, it is advisable to include some spatial characteristics to improve the overall smoothness of the image. However, spatial smoothing in such hyperspectral images is not trivial. The danger of filtering away features of interest is high especially if we do not know where these features lie. Using the same regularization for all feature channels can easily lead to the elimination of the chemical satchels. This is also the reason why no volume smoothing is used here. Volume smoothing tends to eliminate features that appear in only one channel, which is typically the case for peaks in wavelet features. Furthermore, on the one hand it is one of the advantages of the PFS that no manual channel picking has to be performed, but on the other hand one then has to be careful to choose algorithms which do not focus on the channels that contain a lot of noise. The goal of this section is to apply the spatial smoothing approaches from Section 5.6 for this purpose.

6.3 Mockup Mail Bomb

Methods

The first method to improve the smoothness of the segmented image was explained in Algorithm 5.5. Here, a spatial distance between the initial clusters of Algorithm 5.3 is included into the distance matrix calculation. The parameter λ, i.e. the threshold for the overlap that is considered "neighboring", is set to 0.7.

The second method is Algorithm 1. In the first step we use PFS 1 and 6 to calculate a coarse segmentation of the materials. We use the k-means algorithm with the 3 initial cluster centers randomly set and use 100 iterations. The edge preserving smoothing is carried out on all PFS channels separately. The masking is applied on the smoothed coefficients. Some of the noise is mainly salt and pepper noise. Elimination with a median filter works well for these regions. However, there are different kinds of noise that can not be erased with a median filter. For this reason we applied anisotropic diffusion (for details and formulas see [123]). Choosing the contrast and smoothing parameters as well as the number of iterations is not trivial, as the noise is high and the resolution very low. Features of importance can easily be overlooked. We chose to use a contrast of 1.4 and a pre-smoothing-σ of 1.6. After filtering all channels, they are masked with the the medium segment of the k-means segmentation, i.e. the right hand side of Figure 6.24. The masked images are then clustered again by the simplified Chameleon algorithm with Euclidean distance and Average linkage.

The following results were produced with the simplified Chameleon algorithm. They can, however, also be combined with the classical Chameleon. Therefore, the results of the following figures have to be compared to Figure 6.22 to be able to assess the general working of the algorithm.

Results

Figure 6.23 shows some of the resulting levels from the first method that includes a spatial distance. In comparison to the results displayed in Figure 6.22, especially the satchels with the chemicals are much less noisy. In the first clusters they are very smooth and at the same time the wiring within these clusters is still intact. Some of the information, namely the wiring of the ski-card, is not detected until a late and otherwise relatively noisy level (22) but broad structures are present in early clusters.

The second method consists of a multistep image processing algorithm. The first of which, namely the initial three-part segmentation, is displayed on the left

6 Application Examples

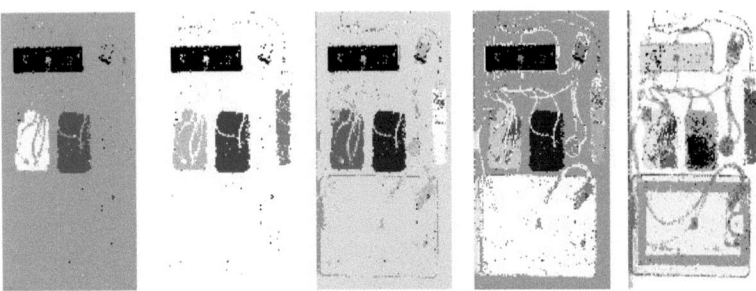

Figure 6.23: The spatial closeness of the initial clusters included into the initial distance matrix, levels 3, 4, 8, 11, and 22.

hand side of Figure 6.24. The right hand side shows the binary mask where only the medium absorbing segment (the gray one from the left hand side) is marked as foreground.

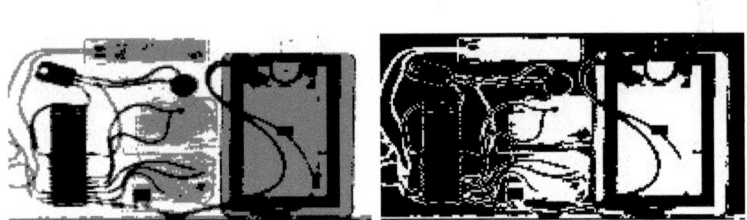

Figure 6.24: Segmentation into three parts with k-means clustering and PFS 1 and 6.

In Figure 6.25 we show some examples channels (unmasked) after the above described smoothing with 50 iterations. The original data is displayed below. The phase slope on the left hand side is filtered very well as well as the wavelet coefficients that contain only said salt and pepper noise. Filtering the wavy shapes from PFS 9 is more difficult, especially as there is also peak information in this channel. All in all the anisotropic diffusion gives good improvement for PFS 1 to 6 and visual improvement that is comparable to a median filter in

6.3 Mockup Mail Bomb

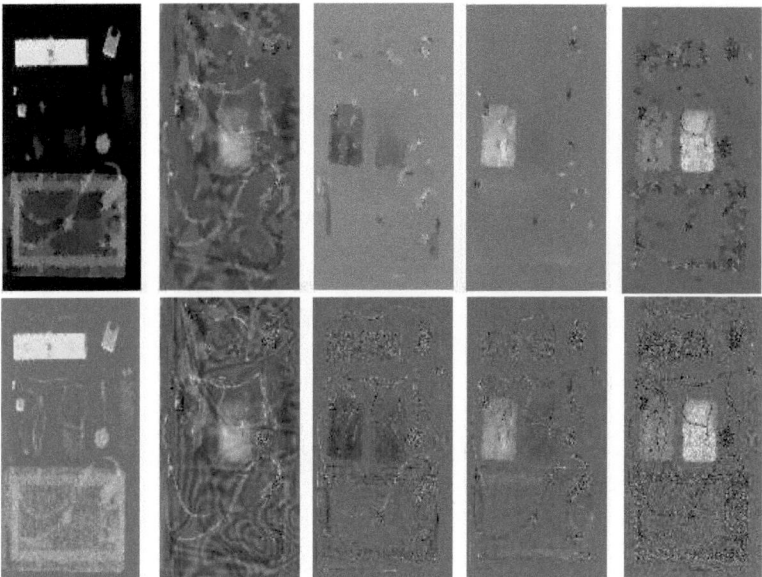

Figure 6.25: PFS 3, 9, 13, 14, and 15 in original (top) and smoothed (bottom).

the later channels. After the smoothing step, the images are masked with the medium segment of the initial segmentation, i.e. the right hand side of Figure 6.24.

Figure 6.26 then shows the first 5 clusters of the simplified Chameleon result on the image processed channels. Naturally the masking of the image produces the first two clusters, i.e. the foreground background segmentation. The next clusters, however, unravel the chemicals both in early levels and in a smooth manner. While the ski-card is shown not to contain much information, the three other components are shown to contain qualitatively different material.

With the aid of this example we have shown that hierarchical clustering is a tool that can be very useful when a segmentation of an unknown number of segments on the basis of various features has to be calculated. Improving such a clustering by including spatial features is a promising approach that improves the overall smoothness while the important features are preserved.

6 Application Examples

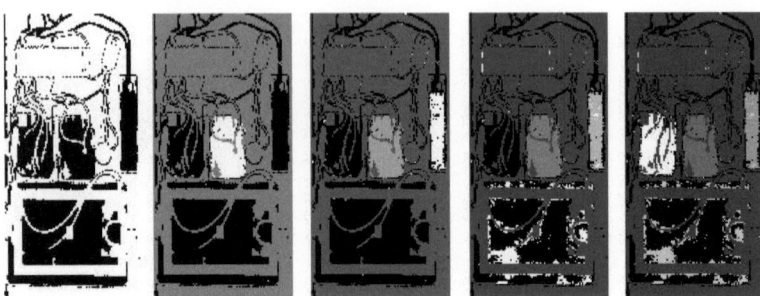

Figure 6.26: Clustering result with spatial inclusion from Algorithm 1 and based on the simplified Chameleon proposed in Algorithm 5.3, levels $1, 2, 3, 4$, and 5.

A next step would be to actually classify the detected components. However, with a correct segmentation such as this one can predict where content of interest might lie and regard single example spectra from these areas instead of classifying all pixels separately.

6.4 Application to Raman Spectra

As a last application example we want to show that various of the proposed methods can not only be used for THz-TDS hyperspectral images but also for hyperspectral images acquired by other techniques. For this purpose we will use one of the techniques introduced in Section 1.1, namely Raman spectroscopy. The image we will use was acquired by B.Heise at the CDL MS-MACH at the Johannes Kepler University, Linz. It is a 70 × 70 pixel measurement of a thin foil of a polymer blend material that contains impurifications. In Figure 6.27 we show a microscopic view of the sampled area. Some impurifications can be seen as bright spots on this image already. On the right hand side of the same figure, we can see an example spectrum taken from this image.

Also in these kinds of spectra, peaks can be observed. Both Raman and THz spectroscopy can be used to identify material by characteristic peaks. However, the peaks here are sharper. Furthermore, Raman spectroscopy is very sensitive. In this measurement for example, typical background shape can be observed which is probably due to autofluorescence. The noise is equally distributed over the measurement and the effect of thermal degradation can appear. This leads to peaks becoming scaled in their height with increasing time. The last effect should be treatable with the cosine distance, for example. The goal in analyzing this measurement is to find inclusions and errors within the material and do so by applying our feature selection and clustering.

Methods

As the spectra are acquired directly in the frequency-domain, we cannot extract PFS 1–5 and lay our emphasis on the remaining coefficients. The wavelet feature selection is applied. We already noted in the previous section that the complex magnitude and the real wavelets display similar information. Therefore, we do not show them both. As the number of peaks is considerably higher in Raman spectra and the peaks are sharper, it seems sensible not just to use the peak height, i.e. the magnitude. The shift of the peaks, that is represented by the phase, is a different criterion that did not work well for THz-TDS measurements but potentially provides additional information for these measurements. Hence, we use the phase and the magnitude of the complex coefficients to extract information about content such as dirt or inclusions.

For the segmentation, we apply classical Chameleon clustering as well as the

6 Application Examples

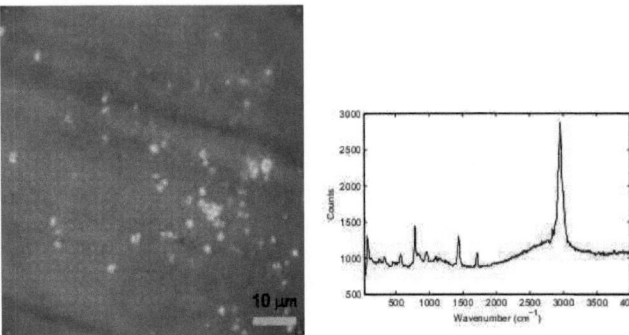

Figure 6.27: Left: Microscopic image of the investigated region with indicated mapping points; Right: An example spectrum. Note: Only the bright spots/impurities are visible in the microscopic image. There are larger patches visible after clustering (Figure 6.30) relating to features at wave number 838cm^{-1} which cannot be seen here.

simplified Chameleon algorithm on both feature sets. For the hierarchical clustering we use Euclidean distance with a complete linkage. For the nearest neighbor graph we chose 100 nearest neighbors and an initial partition of 200.

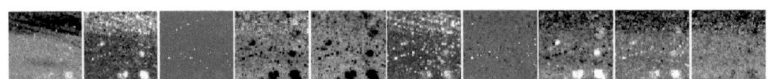

Figure 6.28: Phase of complex PFS 8–17.

Results

Figure 6.28 shows the phase of the complex coefficients and Figure 6.29 shows the magnitude. One can note that the upper and the lower part of the material are separated by both the magnitude and the phase. Nevertheless, the separation of the different feature content does not work as well here as it does in the THz

6.4 Application to Raman Spectra

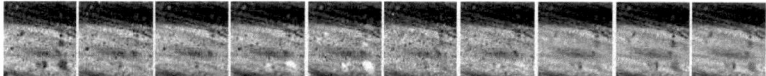

Figure 6.29: Magnitude of complex PFS 8–17.

case (in case of the magnitude PFS 7 to 18 seem to depict more or less the same content). The phase however, depicts very different content from the magnitude. In the lower right areas of the foil, there seem to be circular inclusion that are hardly detected by the magnitude.

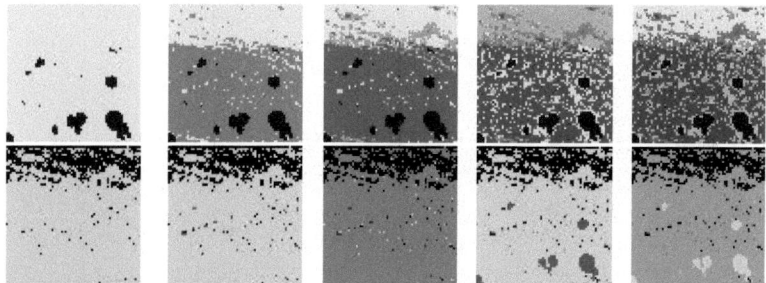

Figure 6.30: Clustering of the phase from Figure 6.28. First five clusters that from the dendrogram retrieved by classical Chameleon clustering (top) and complete linkage simplified Chameleon (bottom).

The results of the clusterings can be seen in Figure 6.30 and 6.31. Although the area in the upper third of the image, where the material is thicker due to a folded region, is detected by both of them, the results between phase and magnitude differ. While the inclusion in the lower right are not detected in the first clusters when the magnitude spectra are used, they are among the first compounds to be found when the phase is used.

As to the performance of the two clustering algorithms: The classical Chameleon detects big clusters in earlier levels but at the same time, higher levels are more prone to error. This result coincides with the result from the previous section.

6 Application Examples

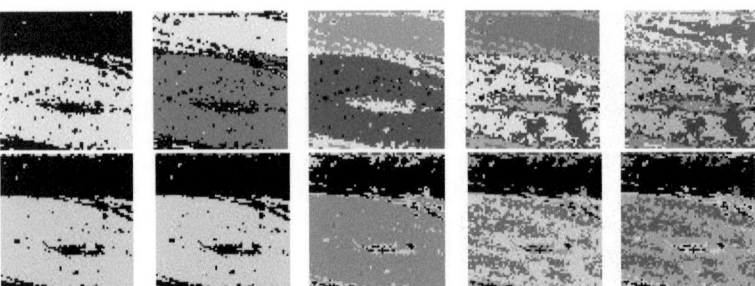

Figure 6.31: Clustering of the magnitude from Figure 6.29. First five clusters that from the dendrogram retrieved by classical Chameleon clustering (top) and complete linkage simplified Chameleon (bottom).

One big advantage of the simplified version is, that different linkage and distance functions can additionally be used and compared, and as the outcome of the classical Chameleon can not be said to always outperform the simplified version, having the choice between both methods improves the chances of getting good results.

Applying the complex PFS coefficients on the Raman hyperspectral images showed promising results. In contrast to the application area of THz-TDS images of chemicals, the phase of these features provided an information gain. The clustering methods were both applicable and yielded good results. Investigating the applicability of this approach on other areas should be a next step

6.5 Conclusion

In this chapter we showed application examples for the methods that were introduced throughout this thesis. With high DR measurements of various chemical compounds as well as a hyperspectral lactose measurement we illustrated that the preprocessing steps introduced the thesis can be used to build and improve THz-TDS databases. It was further shown that the absorption coefficient of such spectra can be represented by the wavelet features and that a segmentation based on this representation is closer to the ground truth than a segmentation based on the full spectrum.

Hyperspectral images with unknown content are typical for "real-world" ap-

6.5 Conclusion

plications such as non-destructive mail inspection. An image from this area was shown to illustrate the robustness of the proposed feature set. The goal of detecting chemical content from such an image can be achieved with these features. The classical Chameleon was applied as well as the introduced simplified Chameleon. Both achieved a valid segmentation. However, the classical Chameleon was more sensitive to noise than the simplified Chameleon with the right choice of linkage and distance function. We successfully applied two methods that include spatial information to further improve the clustering result.

Furthermore, we showed that the feature extraction as well as the segmentation tools can be adapted to be used for other hyperspectral imaging areas on the example of a Raman measurement.

6 Application Examples

7 Conclusion

The problem that formed the basis of this thesis was to find a representation for hyperspectral Terahertz images. This representation was supposed to be usable to detect different materials with a special focus on the detection of chemicals. Therein as much information as possible about the specific shape of these spectra was to be considered while at the same time it should be adaptable to various applications. Furthermore, setting a lot of parameters was to be avoided. For these reasons we proposed to apply a course of action consisting of three steps: Preprocessing, feature reduction and segmentation — as shown in Figure 1.3 in the introduction (Chapter 1). These three steps, hence, built the thread running through this work and which is represented in its main chapters.

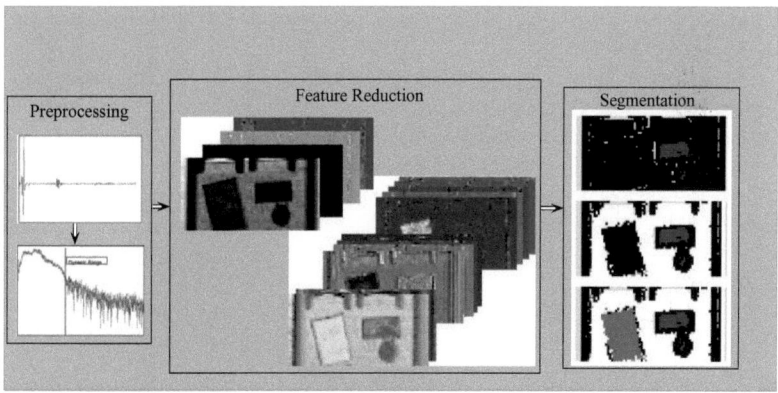

Figure 7.1: Illustration of procedure from Figure 1.3 on image from Figure 1.1. In the middle PFS features 1–4 and 6–15 are shown.

7 Conclusion

In Figure 1.1 in the introduction we have shown some channels of a hyperspectral image to illustrate this problem. The same image will now be used again in Figure 7.1 to illustrate the usefulness of the methods developed in this thesis.

The left hand side of the figure illustrates preprocessing steps that were the focus of Chapter 2. Therein, we analyzed and applied standard procedures and additionally introduced new methods for THz-TDS spectra. Those were echo removal in the time-domain, dynamic range determination in the frequency-domain and baseline correction. The first two are shown in Figure 7.1. Potential application for these preprocessing steps are, however, not just in hyperspectral imaging but also in the analysis of single spectra, and can be used for building reliably data bases of peak positions and depth of chemical compounds. This was illustrated by some application measurements in Chapter 6.

In Figure 1.1 we showed 20 of the 3200 initially measured channels, they were by no means representative of the spectral content of the measurement. The images shown in the middle of Figure 7.1, however, show 15 features that represent time-domain and phase characteristics (top) and the full dynamic range of the magnitude spectrum (bottom), i.e. reduce the initial 3200 feature down to 15 without any previous selection of regions of interest. In Excursus 2 we introduced this feature set, that is based on features that are often used for these spectra as well as new wavelet based sets representing the magnitude spectrum. They were introduced in Chapter 4 and are generated by real wavelet coefficients as well as complex wavelet coefficients. Their adaptability to different application scenarios was shown in Chapter 6.

To evaluate if this feature reduction is valid for THz-TDS spectra, it does not suffice to show their applicability on some example measurements. We therefore did an extensive evaluation of the representativity of the PFS with simulated spectra. The basis for this evaluation was a simulation scheme that we introduced in Chapter 3. This simulation scheme is particularly designed to build an arbitrary big data base of THz-TDS spectra with varying number, depth and position of peaks. It takes the specific shape as well as the noise of these spectra into account.

The segmentation was done with hierarchical clustering. In Chapter 5 we discussed different hierarchical clustering algorithms. Different parameters for the classical algorithm were analyzed. To cope with high volume data we discussed Karypis Chameleon algorithm. We propose alterations to this algorithm to make it stable to inversions and to be able to use the classical algorithm's parameters. Furthermore we discussed two approaches that are used to include spatial features into the hyperspectral clustering. Finally an evaluation scheme was introduced.

In Figure 7.1 the right hand side shows such a segmentation result. We show different clustering levels, this possibility is one of the main advantages of hierarchical clustering. Further applications of all the methods including the spatial clustering and the evaluation on real-world data could be seen in Chapter 6.

Furthermore the wavelet based feature reduction as well as the hierarchical clustering was applied on Raman spectra to show these methods are applicable to different hyperspectral data.

With all these methods we are now able to analyze hyperspectral THz-TDS images in a comprehensive and reproducible manner. Most of the methods need no, or only few parameters. This is an important point, especially for the methods that are applied on the single spectra as it is hardly feasible to smooth, baseline correct or otherwise alter the spectra individually.

All in all the methods are designed especially to detect chemical compounds in hyperspectral THz-TDS dry air measurements at room temperature. As to the generality of these methods, the preprocessing is THz specific, but can also be used to improve individual THz-TDS spectra. While the feature reduction is applicable on other time-series measurements with comparably broad characteristics and the segmentation can be used for all types of hyperspectral images.

7.1 Future Research

Although the goal of this work was to give a self contained course of action to process THz-TDS data, there always remain things to be done and topics to be further investigated. We will now name some.

Classification: It was already said that no final classification is done in this work. In specific applications one does not stop at segmenting an image but needs to proceed to form some qualitative statement about the segments. In drug detection applications, for example, one would want to know, if there was a drug detected, yes or no, and if yes, which drug. This means that a subsequent classification of the image content should be executed.

However, a good segmentation leads to a faster and more reliable classification result. In Figure 7.2 we show the segmentation of the example image in 5 clusters again on the left hand side and the example spectra for the respective segments on the right hand side. Instead of classifying all 2888 spectra, one would have

7 Conclusion

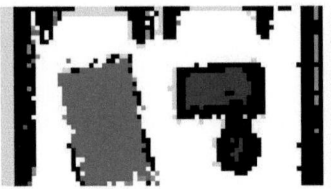 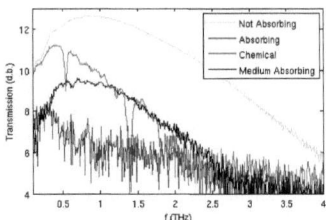

Figure 7.2: Left: Segmentation of example image to 5 regions; Right: Example spectra from each of the regions.

to classify only these 5 regions of interest. Further research should therefore aim at combining the given segmentation tools as well as the feature reduction with classification algorithms.

Improving the Spatial Information Inclusion: Including spatial information into hyperspectral data organization is not at all trivial. Because the quality of many images is not good, filtering noise out often leads to the elimination of important information. We showed two possibilities to include spatial information nevertheless. However, the resulting images are still quite noisy. Further research should also be done in this area.

Further Investigation of Usability of Phase of Complex Wavelet Coefficients:
Furthermore, in Chapter 4 we introduced a feature set that is based on the complex wavelet transform. Most of the applications we showed only used the magnitude spectrum, as the phase was too sensitive to noise. In the application to hyperspectral Raman data, the phase spectrum of these coefficients unraveled different information than the magnitude. The application of the phase of these coefficients should therefore be further investigated.

Implementations

All algorithms were implemented in C++ and / or Matlab by the author.

Johannes Kepler University: Most of the preprocessing algorithms as well as the feature reduction and the clustering were included into a graphical user interface (GUI) by Ulrich Brandstätter from the Johannes Kepler University Linz. A screenshot of this GUI is shown in Figure 7.3. This GUI can be used and is

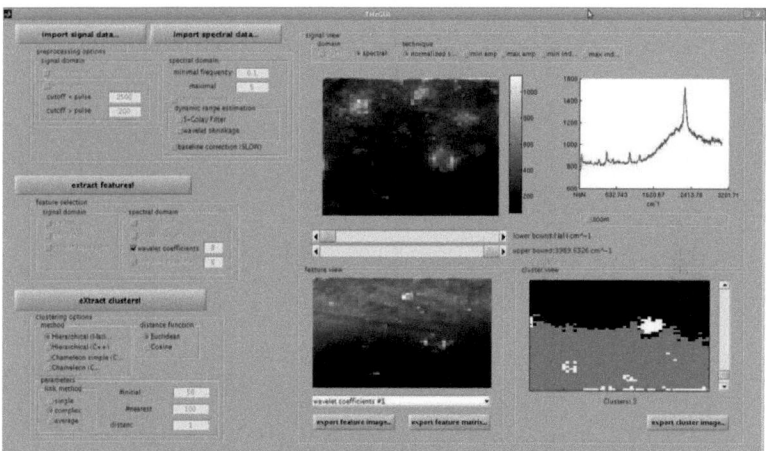

Figure 7.3: GUI for processing hyperspectral THz images

used for different hyperspectral images. The preprocessing that was especially developed for the THz region has limited applicability to other spectral regions.

7 Conclusion

However, feature reduction and data segmentation can be applied on arbitrary hyperspectral images.

7.1 Future Research

Fraunhofer ITWM: The authors implementation of simplified Chameleon as well as the classical hierarchical clustering were adapted to be usable with the ToolIP framework of the image processing department of the Fraunhofer ITWM by two students — Maximilian Leitheiser and Razvan Turtoi — supervised by the author. A snapshot of the implemented nodes in shown in Figure 7.4. One

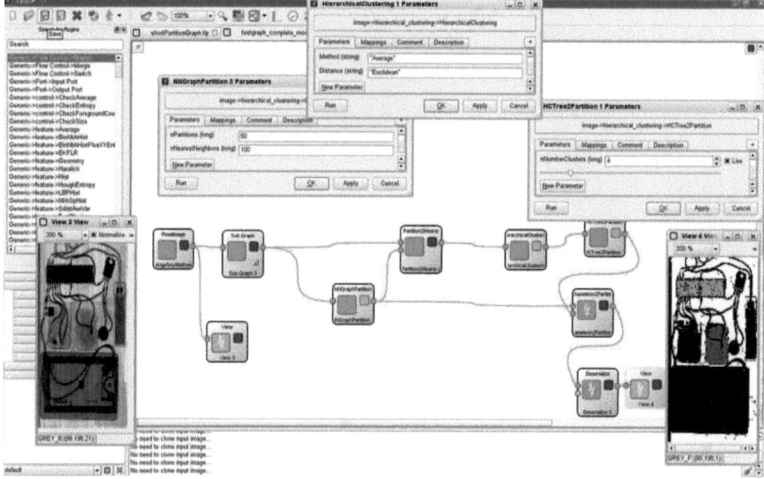

Figure 7.4: Five plugins for the usage of hierarchical clustering in the ToolIP framework

can see the different parameters that can be set. These plugins are not restricted to hyperspectral data only feature channels are needed.

155

7 Conclusion

Parts of this Work Previously Published In:

[1] H. Stephani, M. Herrmann, K. Wiesauer, S. Katletz, and B. Heise. Enhancing the interpretability of terahertz data through unsupervised classification. In *2009 IMEKO World Congress*, page 637, September 2009.

[2] H. Stephani, J. Jonuscheit, C. Robine, and B. Heise. Automatically detecting peaks in terahertz time-domain spectroscopy. In *2010 20th International Conference on Pattern Recognition (ICPR)*, pages 4468–4471. IEEE, August 2010.

[3] H. Stephani, M. Herrmann, F. Bauer, and B. Heise. Wavelet-based dimensionality reduction for hyperspectral terahertz imaging. *International Journal of Terahertz Science and Technology*, 3(3):117–129, October 2010.

[4] H. Stephani, B. Heise, K. Wiesauer, S. Katletz, D. Molter, J. Jonuscheit, and R. Beigang. A feature set for enhanced automatic segmentation of hyperspectral terahertz images. In *2011 Irish Machine Vision and Image Processing Conference (IMVIP)*, pages 117–122. IEEE, September 2011.

[5] H. Stephani, C. Robine, K. Wiesauer, S. Katletz, and B. Heise. Approaches in automatic hyperspectral image segmentation. In *4th International Workshop on Terahertz Technology*, March 2010.

Parts of this Work Previously Published In:

Bibliography

[6] D. Landgrebe. Hyperspectral image data analysis. *Signal Processing Magazine, IEEE*, 19(1):17–28, 2002.

[7] C.I. Chang. *Hyperspectral Data Exploitation: Theory and Applications*. Wiley-Blackwell, 2007.

[8] H. Grahn and P. Geladi. *Techniques and Applications of Hyperspectral Image Analysis*. Wiley, 2007.

[9] M. Stoeckli, P. Chaurand, D.E. Hallahan, and R.M. Caprioli. Imaging mass spectrometry: a new technology for the analysis of protein expression in mammalian tissues. *Nature Medicine*, 7(4):493–496, 2001.

[10] J.W. Handley. *Time Frequency Analysis Techniques in Terahertz Pulsed Imaging*. PhD thesis, University of Leeds, 2003.

[11] E. Smith and G. Dent. *Modern Raman Spectroscopy: a Practical Approach*. Wiley, 2005.

[12] E.C. Le Ru and P.G. Etchegoin. *Principles of Surface-Enhanced Raman Spectroscopy: and Related Plasmonic Effects*. Elsevier Science, 2008.

[13] H. Cen and Y. He. Theory and application of near infrared reflectance spectroscopy in determination of food quality. *Trends in Food Science & Technology*, 18(2):72–83, 2007.

[14] S. Bengtsson, T. Berglöf, and H. Kylin. Near infrared reflectance spectroscopy as a tool to predict pesticide sorption in soil. *Bulletin of Environmental Contamination and Toxicology*, 78(5):295–298, 2007.

[15] M. Blanco, J. Coello, H. Iturriaga, S. Maspoch, and C. De La Pezuela. Near-infrared spectroscopy in the pharmaceutical industry. critical review. *Analyst*, 123(8):135R–150R, 1998.

Bibliography

[16] Y. Roggo, P. Chalus, L. Maurer, C. Lema-Martinez, A. Edmond, and N. Jent. A review of near infrared spectroscopy and chemometrics in pharmaceutical technologies. *Journal of Pharmaceutical and Biomedical Analysis*, 44(3):683–700, 2007.

[17] P.M. Chu, F.R. Guenther, G.C. Rhoderick, and W.J. Lafferty. The nist quantitative infrared database. *Journal of Research - National Institute of Standards and Technology*, 104:59–82, 1999.

[18] S. Wartewig and R.H.H. Neubert. Pharmaceutical applications of mid-ir and raman spectroscopy. *Advanced Drug Delivery Reviews*, 57(8):1144–1170, 2005.

[19] S. Gorenflo. *A Comprehensive Study of Macromolecules in Composites Using Broadband Terahertz Spectroscopy*. PhD thesis, Universitaetsbibliothek Freiburg, 2006.

[20] L. Maurer. *Near Infrared Spectroscopy/Imaging and Terahertz Spectroscopy/Imaging for the Analysis of Solid Dosage Forms*. PhD thesis, Universität Basel, 2008.

[21] M. Tonouchi. Cutting-edge terahertz technology. *Nature Photonics*, 1(2):97–105, 2007.

[22] C. Sirtori. Bridge for the terahertz gap. *Nature*, 417(6885):132–3, 2002.

[23] D.E. Spence, P.N. Kean, and W. Sibbett. 60-fsec pulse generation from a self-mode-locked ti: Sapphire laser. *Optics Letters*, 16(1):42–44, 1991.

[24] C.J. Strachan, P.F. Taday, D.A. Newnham, K.C. Gordon, J.A. Zeitler, M. Pepper, and T. Rades. Using terahertz pulsed spectroscopy to quantify pharmaceutical polymorphism and crystallinity. *Journal of Pharmaceutical Sciences*, 94(4):837–846, 2005.

[25] L.S. Rothman, I.E. Gordon, A. Barbe, D.C. Benner, P.F. Bernath, M. Birk, V. Boudon, L.R. Brown, A. Campargue, J.P. Champion, et al. The hitran 2008 molecular spectroscopic database. *Journal of Quantitative Spectroscopy and Radiative Transfer*, 110(9-10):533–572, 2009.

Bibliography

[26] Y. Wang, Z. Zhao, Z. Chen, Y. Zhang, L. Zhang, and K. Kang. Suppression of spectral interferences due to water-vapor rotational transitions in terahertz time-domain spectroscopy. *Optics Letters*, 33(12):1354–1356, 2008.

[27] P.U. Jepsen and B.M. Fischer. Dynamic range in terahertz time-domain transmission and reflection spectroscopy. *Optics Letters*, 30(1):29–31, 2005.

[28] W.L. Chan, J. Deibel, and D.M. Mittleman. Imaging with terahertz radiation. *Reports on Progress in Physics*, 70:1325–1379, 2007.

[29] B. Pradarutti, G. Matthaeus, S. Riehemann, G. Notni, S. Nolte, and A. Tuennermann. Advanced analysis concepts for terahertz time domain imaging. *Optics Communications*, 279(2):248–254, 2007.

[30] F.J. Harris. On the use of windows for harmonic analysis with the discrete fourier transform. *Proceedings of the IEEE*, 66(1):51–83, 1978.

[31] A. Nuttall. Some windows with very good sidelobe behavior. *IEEE Transactions on Acoustics, Speech and Signal Processing*, 29(1):84–91, 1981.

[32] J.G. Proakis and D.G. Manolakis. *Digital Signal Processing: Principles, Algorithms, and Applications*. Prentice-Hall, Inc.Upper Saddle River, NJ, USA, 1996.

[33] S. Mallat. *A Wavelet Tour of Signal Processing*. Academic Press, 1999.

[34] I. Daubechies. *Ten Lectures on Wavelets*. Society for Industrial and Applied Mathematics (SIAM), 1992.

[35] K. Bredies and D. Lorenz. *Mathematische Bildverarbeitung: Einführung in Grundlagen und moderne Theorie*. Vieweg und Teubner Verlag, 2011.

[36] D. Donoho, A. Maleki, and M. Shaharam. Wavelab 850. *Software Toolkit for Time-Frequency Analysis*, 2006.

[37] T.D. Dorney, R.G. Baraniuk, and D.M. Mittleman. Material parameter estimation with terahertz time-domain spectroscopy. *JOSA A*, 18(7):1562–1571, 2001.

Bibliography

[38] F. Huang, J.F. Federici, and D. Gary. Determining thickness independently from optical constants by use of ultrafast light. *Optics Letters*, 29(20):2435–2437, 2004.

[39] C. Taswell. The what, how, and why of wavelet shrinkage denoising. *Computing in Science & Engineering*, 2(3):12–19, 2000.

[40] M. Golay and A. Savitzky. Smoothing and differentiation of data by simplified least square procedures. *Analytical Chemistry*, 36:1627–1639, 1964.

[41] H. Mark and J. Workman. *Chemometrics in Spectroscopy*. Academic Press, 2007.

[42] J. Chen, P. Jonsson, M. Tamura, Z. Gu, B. Matsushita, and L. Eklundh. A simple method for reconstructing a high-quality ndvi time-series data set based on the savitzky-golay filter. *Remote Sensing of Environment*, 91(3-4):332–344, 2004.

[43] V.J. Barclay, R.F. Bonner, and I.P. Hamilton. Application of wavelet transforms to experimental spectra: Smoothing, denoising, and data set compression. *Analytical Chemistry*, 69(1):78–90, 1997.

[44] N. Laman, S. Sree Harsha, D. Grischkowsky, and J.S. Melinger. 7 ghz resolution waveguide thz spectroscopy of explosives related solids showing new features. *Optics Express*, 16(6):4094–4105, 2008.

[45] M. Blanco-Velasco, B. Weng, and K.E. Barner. Ecg signal denoising and baseline wander correction based on the empirical mode decomposition. *Computers in Biology and Medicine*, 38(1):1–13, 2008.

[46] A.C. Sauve and T.P. Speed. Normalization, baseline correction and alignment of high-throughput mass spectrometry data. *Proceedings Gensips*, 2004.

[47] S. Wold, H. Antti, F. Lindgren, and J. Öhman. Orthogonal signal correction of near-infrared spectra. *Chemometrics and Intelligent Laboratory Systems*, 44(1-2):175–185, 1998.

[48] Y. Xi and D.M. Rocke. Baseline correction for nmr spectroscopic metabolomics data analysis. *BMC Bioinformatics*, 9(1):324, 2008.

[49] Y. Ueno and K. Ajito. Analytical terahertz spectroscopy. *Analytical Sciences*, 24(2):185–192, 2008.

[50] Z.S. Piao, M. Tani, and K. Sakai. Carrier dynamics and thz radiation in biased semiconductor structures. In *Society of Photo-Optical Instrumentation Engineers (SPIE) Conference Series*, volume 3617, pages 49–56, 1999.

[51] L. Duvillaret, F. Garet, J.F. Roux, and J.L. Coutaz. Analytical modeling and optimization of terahertz time-domain spectroscopy experiments using photoswitches as antennas. *IEEE Journal on Selected Topics in Quantum Electronics*, 7(4):615–623, 2001.

[52] A. Bonvalet and M. Joffre. Terahertz femtosecond pulses. *Femtosecond Laser Pulses: Principles and Experiments*, 1995.

[53] L. Duvillaret, F. Garet, and J.L. Coutaz. Influence of noise on the characterization of materials by terahertz time-domain spectroscopy. *Journal of the Optical Society of America B*, 17(3):452–461, 2000.

[54] H. Stöcker. *Taschenbuch Mathematischer Formeln und Moderner Verfahren*. Harri Deutsch Verlag, 2009.

[55] A.V. Chechkin and V.Y. Gonchar. Fractional brownian motion approximation based on fractional integration of a white noise. *Chaos, Solitons and Fractals*, 12(2):391–398, 2001.

[56] A.V. Chechkin and V.Y. Gonchar. Fractional brownian motion approximation based on fractional integration of a white noise. *Chaos, Solitons & Fractals*, 12(2):391–398, 2001.

[57] B.M. Fischer. Chemische analytik und bildgebung mit gepulster terahertzstrahlung. *www.analytik-news.de*, 2009.

[58] P.F. Taday. Applications of terahertz spectroscopy to pharmaceutical sciences. *Philosophical Transactions A*, 362(1815):351–364, 2004.

[59] I.V. Hertel and C.P. Schulz. *Atome, Moleküle und Optische Physik 2: Moleküle und Photonen-Spektroskopie und Streuphysik*. Springer, 2010.

Bibliography

[60] J.E. Huheey, E.A. Keiter, and R. Keiter. *Anorganische Chemie: Prinzipien von Struktur und Reaktivität*, chapter Koordinationsverbindungen: Bindungstheorie, Spektren und Magnetismus. Walter de Gruyter, 2003.

[61] J.F. Federici, B. Schulkin, F. Huang, D. Gary, R. Barat, F. Oliveira, and D. Zimdars. Thz imaging and sensing for security applications – explosives, weapons and drugs. *Semiconductor Science Technology*, 20:266–280, 2005.

[62] M. Steinbach, L. Ertoes, and V. Kumar. The challenges of clustering high dimensional data. *New Directions in Statistical Physics: Econophysics, Bioinformatics, and Pattern Recognition*, pages 273–307, 2004.

[63] Y. Saeys, I. Inza, and P. Larrañaga. A review of feature selection techniques in bioinformatics. *Bioinformatics*, 23(19):2507, 2007.

[64] B. Ferguson, S. Wang, D. Gray, D. Abbott, and X.C. Zhang. Terahertz imaging of biological tissue using a chirped probe pulse. In *Society of Photo-Optical Instrumentation Engineers (SPIE) Conference Series*, volume 4591, pages 172–84, 2001.

[65] J.B. Jackson, M. Mourou, J.F. Whitaker, I.N. Duling III, S.L. Williamson, M. Menu, and G.A. Mourou. Terahertz imaging for non-destructive evaluation of mural paintings. *Optics Communications*, 281(4):527–532, 2008.

[66] M. Herrmann, R. Fukasawa, and O. Morikawa. Terahertz imaging. In K.Sakai, editor, *Terahertz Optoelectronics, Topics Appl.Phys.*, volume 97, pages 331 – 382. Springer-Verlag Berlin Heidelberg, 2005.

[67] K. Kawase, Y. Ogawa, Y. Watanabe, and H. Inoue. Non-destructive terahertz imaging of illicit drugs using spectral fingerprints. *Optics Express*, 11(20):2549–2554, 2003.

[68] A.K. Jain and R.C. Dubes. *Algorithms for Clustering Data*. Prentice-Hall, New Jersey, 1988.

[69] I. Guyon and A. Elisseeff. An introduction to variable and feature selection. *Journal of Machine Learning Research*, 3:1157–1182, 2003.

[70] A. Jain and D. Zongker. Feature selection: Evaluation, application, and small sample performance. *IEEE Transactions on Pattern Analysis and Machine Intelligence*, 19(2), 1997.

Bibliography

[71] M.J. Mendenhall and E. Merenyi. Relevance-based feature extraction for hyperspectral images. *IEEE Transactions on Neural Networks*, 19(4):658–672, 2008.

[72] J.G. Dy and C.E. Brodley. Feature selection for unsupervised learning. *Journal of Machine Learning Research*, 5:845–889, 2004.

[73] J. Friedman, T. Hastie, and R. Tibshirani. *The Elements of Statistical Learning*. Springer Series in Statistics, 2009.

[74] R. Aebersold and M. Mann. Mass spectrometry-based proteomics. *Nature*, 422(6928):198–207, 2003.

[75] D.M. Mittleman, M. Gupta, R. Neelamani, R.G. Baraniuk, J.V. Rudd, and M. Koch. Recent advances in terahertz imaging. *Applied Physics B: Lasers and Optics*, 68(6):1085–1094, 1999.

[76] H. Ding, G. Trajcevski, P. Scheuermann, X. Wang, and E. Keogh. Querying and mining of time series data: Experimental comparison of representations and distance measures. *Proceedings of the VLDB Endowment*, 1(2):1542–1552, 2008.

[77] H. Zhao, P.C. Yuen, and J.T. Kwok. A novel incremental principal component analysis and its application for face recognition. *IEEE Transactions on Systems, Man, and Cybernetics, Part B: Cybernetics*, 36(4):873–886, 2006.

[78] Y. Chen, S. Huang, and E. Pickwell-MacPherson. Frequency-wavelet domain deconvolution for terahertz reflection imaging and spectroscopy. *Optics Express*, 18(2):1177–1190, 2010.

[79] D.M. Mittleman, R.H. Jacobsen, and M.C. Nuss. T-ray imaging. *IEEE Journal of Selected Topics in Quantum Electronics*, 2(3):679–692, 1996.

[80] W. Zhao and C.E. Davis. Swarm intelligence based wavelet coefficient feature selection for mass spectral classification: an application to proteomics data. *Analytica Chimica Acta*, 651(1):15, 2009.

[81] R.K.H. Galvao and T. Yoneyama. A competitive wavelet network for signal clustering. *IEEE Transactions on Systems, Man, and Cybernetics, Part B*, 34(2):1282–1288, 2004.

Bibliography

[82] H. Zhang, T.B. Ho, Y. Zhang, and M.S. Lin. Unsupervised feature extraction for time series clustering using orthogonal wavelet transform. *Informatica*, 30:305–319, 2006.

[83] N. Kingsbury. Complex wavelets for shift invariant analysis and filtering of signals. *Applied and Computational Harmonic Analysis*, 10:234–253, 2001.

[84] M.J. Mendenhall and E. Merényi. Relevance-based feature extraction from hyperspectral images in the complex wavelet domain. In *2006 IEEE Workshop on Adaptive and Learning Systems*, pages 24–29. IEEE, 2006.

[85] I.W. Selesnick, R.G. Baraniuk, and N.G. Kingsbury. The dual-tree complex wavelet transform. *IEEE Signal Processing Magazine*, 22(6):123–151, 2005.

[86] J.L. Rodgers and W.A. Nicewander. Thirteen ways to look at the correlation coefficient. *American Statistician*, pages 59–66, 1988.

[87] J. Bergh, F. Ekstedt, and M. Lindberg. *Wavelets mit Anwendungen in Signal-und Bildbearbeitung*. Springer Verlag, 2007.

[88] D.M. Mittleman, J. Cunningham, M.C. Nuss, and M. Geva. Noncontact semiconductor wafer characterization with the terahertz hall effect. *Applied Physics Letters*, 71:16–18, 1997.

[89] D. Arthur and D. Vassilvitskii. How slow is the k-means method? In *SCG '06: Proceedings of the Twenty-Second Annual Symposium on Computational Geometry*, pages 144–153, New York, NY, USA, 2006. ACM.

[90] C.M. Bishop. *Pattern Recognition and Machine Learning*. Springer-Verlag New York, Inc., Secaucus, NJ, USA, 2006.

[91] T. Kohonen. Self-organization of very large document collections: State of the art. In L.Niklasson, M.Bodén, and T.Ziemke, editors, *Proceedings of ICANN98, the 8th International Conference on Artificial Neural Networks*, volume 1, pages 65–74. Springer, London, 1998.

[92] P. Berkhin. A survey of clustering data mining techniques. *Grouping Multidimensional Data: Recent Advances in Clustering*, pages 25–71, 2006.

[93] L. Fahrmeir, W. Brachinger, A. Hamerle, and G. Tutz. *Multivariate Statistische Verfahren*. de Gruyter, 1996.

Bibliography

[94] D. Steinhausen and K. Langer. *Clusteranalyse*. de Gruyter, 1977.

[95] J. Ward and H. Joe. Hierarchical grouping to optimize an objective function. *Journal of the American Statistical Association*, 58(301):pp.236–244, 1963.

[96] R. Xu and D. Wunsch. Survey of clustering algorithms. *IEEE Transactions on Neural Networks*, 16(3):645–678, 2005.

[97] T. Korenius, J. Laurikkala, and M. Juhola. On principal component analysis, cosine and euclidean measures in information retrieval. *Information Sciences*, 177(22):4893–4905, 2007.

[98] G.N. Lance and W.T. Williams. A general theory of classificatory sorting strategies: 1.hierarchical systems. *The Computer Journal*, 9(4):373, 1967.

[99] T.W. Liao. Clustering of time series data - a survey. *Pattern Recognition*, 38:1857 – 1874, 2005.

[100] T. Zhang, R. Ramakrishnan, and M. Livny. Birch: An efficient data clustering method for very large databases. *SIGMOD Rec.*, 25(2):103–114, 1996.

[101] S. Guha, R. Rastogi, and K. Shim. Cure: An efficient clustering algorithm for large databases. In *ACM SIGMOD Record*, volume 27, pages 73–84. ACM, 1998.

[102] G. Karypis, E.H. Han, and V. Kumar. Chameleon: Hierarchical clustering using dynamic modeling. *Computer*, 32(8):68–75, 1999.

[103] D. Jungnickel. *Graphs, Networks and Algorithms*. Springer Verlag, 2007.

[104] M.B. Kennel. KDTREE 2: Fortran 95 and C++ Software to Efficiently Search for Near Neighbors in a Multi-Dimensional Euclidean Space. *ArXiv Physics e-Prints*, August 2004.

[105] G. Karypis and V. Kumar. A fast and high quality multilevel scheme for partitioning irregular graphs. *SIAM Journal on Scientific Computing*, 20(1):359, 1999.

Bibliography

[106] G. Karypis and V. Kumar. A fast and high quality multilevel scheme for partitioning irregular graphs. *SIAM Journal on Scientific Computing*, 20(1):359, 1999.

[107] S. Huang, E. Aubanel, and V. Bhavsar. Mesh partitioners for computational grids: a comparison. *Computational Science and Its Applications ICCSA 2003*, pages 985–985, 2003.

[108] G. Karypis. Cluto-a clustering toolkit. Technical report, Minnesota Univ Minneapolis Dept of Computer Science, 2002.

[109] A. Plaza, J.A. Benediktsson, J.W. Boardman, J. Brazile, L. Bruzzone, G. Camps-Valls, J. Chanussot, M. Fauvel, P. Gamba, A. Gualtieri, et al. Recent advances in techniques for hyperspectral image processing. *Remote Sensing of Environment*, 113:S110–S122, 2009.

[110] M. Lennon, G. Mercier, and L. Hubert-Moy. Nonlinear filtering of hyperspectral images with anisotropic diffusion. In *Geoscience and Remote Sensing Symposium, 2002.IGARSS'02.2002 IEEE International*, volume 4, pages 2477–2479. IEEE, 2002.

[111] J. Martin-Herrero and M. Ferreiro-Arman. Tensor-driven hyperspectral denoising: A strong link for classification chains? In *Pattern Recognition (ICPR), 2010 20th International Conference on*, pages 2820–2823. IEEE, 2010.

[112] J. Weickert. *Anisotropic Diffusion in Image Processing*. Teubner Verlag, Stuttgart, 1998.

[113] C. Tomasi and R. Manduchi. Bilateral filtering for gray and color images. In *Proceedings of the Sixth International Conference on Computer Vision*, page 839. IEEE Computer Society, 1998.

[114] M.Halikidi, Y.Batistakis, and M.Vazirgiannis. On clustering validation techniques. *Journal of Intelligent Information Systems*, 17(2/3):107–145, 2001.

[115] J.C. Dunn. Well-separated clusters and optimal fuzzy partitions. *Journal of Cybernetics*, 4(1):95–104, 1974.

Bibliography

[116] M. Meila. Comparing clusterings – an information based distance. *Journal of Multivariate Analysis*, 98(5):873–895, 2007.

[117] W.M. Rand. Objective criteria for the evaluation of clustering methods. *Journal of the American Statistical association*, 66(336):846–850, 1971.

[118] N. Jardine and C.J. van Rijsbergen. The use of hierarchic clustering in information retrieval. *Information Storage and Retrieval*, 7(5):217–240, 1971.

[119] B. Larsen and C. Aone. Fast and effective text mining using linear-time document clustering. In *Proceedings of the Fifth ACM SIGKDD International Conference on Knowledge Discovery and Data Mining*, pages 16–22. ACM, 1999.

[120] Y. Zhao and G. Karypis. Evaluation of hierarchical clustering algorithms for document datasets. In *Proceedings of the Eleventh International Conference on Information and Knowledge Management*, pages 515–524. ACM, 2002.

[121] NICT and RIKEN. The terahertz database.

[122] E.R. Brown, J.E. Bjarnason, A.M. Fedor, and T.M. Korter. On the strong and narrow absorption signature in lactose at 0.53 thz. *Applied Physics Letters*, 90:061908, 2007.

[123] J. Weickert, B.M.T.H. Romeny, and M.A. Viergever. Efficient and reliable schemes for nonlinear diffusion filtering. *IEEE Transactions on Image Processing*, 7(3):398–410, 1998.

i want morebooks!

Buy your books fast and straightforward online - at one of world's fastest growing online book stores! Environmentally sound due to Print-on-Demand technologies.

Buy your books online at
www.get-morebooks.com

Kaufen Sie Ihre Bücher schnell und unkompliziert online – auf einer der am schnellsten wachsenden Buchhandelsplattformen weltweit! Dank Print-On-Demand umwelt- und ressourcenschonend produziert.

Bücher schneller online kaufen
www.morebooks.de

VDM Verlagsservicegesellschaft mbH
Heinrich-Böcking-Str. 6-8   Telefon: +49 681 3720 174   info@vdm-vsg.de
D - 66121 Saarbrücken   Telefax: +49 681 3720 1749   www.vdm-vsg.de

Printed by Books on Demand GmbH, Norderstedt / Germany